① *Headline*: Section 2.2.3.

Short Hermann–Mauguin symbol	Schoenflies symbol	Crystal class (Point group)	Crystal system
(Section 2.2.4 and Chapter 12.2)	(Chapters 12.1 and 12.2)	(Section 10.1.1 and Chapter 12.1)	(Section 2.1.2)

②

Number of space group	Full Hermann–Mauguin symbol	Patterson symmetry
[Same as in *IT* (1952)]	(Section 2.2.4 and Chapter 12.3)	(Section 2.2.5)

③ *Space-group diagrams*, consisting of one or several projections of the symmetry elements and one illustration of a set of equivalent points in general position. The numbers and types of the diagrams depend on the crystal system. The diagrams and their axes are described in Section 2.2.6; the graphical symbols of symmetry elements are listed in Chapter 1.4.

For monoclinic space groups see Section 2.2.16; for orthorhombic settings see Section 2.2.6.4.

④ *Origin* of the unit cell: Section 2.2.7. The site symmetry of the origin and its location with respect to the symmetry elements are given.

⑤ *Asymmetric unit*: Section 2.2.8. One choice of asymmetric unit is given.

⑥ *Symmetry operations*: Section 2.2.9 and Part 11. For each point $\tilde{x}, \tilde{y}, \tilde{z}$ of the general position that symmetry operation is listed which transforms the initial point x, y, z into the point under consideration. The symbol describes the nature of the operation, its glide or screw component (given between parentheses), if present, and the location of the corresponding symmetry element.

The symmetry operations are numbered in the same way as the corresponding coordinate triplets of the general position. For centred space groups the same numbering is applied in each block, *e.g.* under 'For $(\frac{1}{2}, \frac{1}{2}, 0)+$ set'.

[Continued on inside back cover]

Note: The chapter and section numbers refer to Volume A and, in part, to this *Teaching Edition*.

EXPLANATION OF THE SPACE-GROUP DATA

INTERNATIONAL TABLES

FOR

CRYSTALLOGRAPHY

Brief Teaching Edition of
Volume A
SPACE-GROUP SYMMETRY

Edited by
THEO HAHN

Fifth Edition

Published for
THE INTERNATIONAL UNION OF CRYSTALLOGRAPHY

by

John Wiley & Sons, Ltd

2014

A C.I.P. Catalogue record for this book
is available from the Library of Congress
ISBN 978-0-470-68911-0 (acid-free paper)

First published in 1985
Second, revised edition 1988
Reprinted with corrections 1989
Third, revised and enlarged edition 1993
Fourth, revised and enlarged edition 1996
Reprinted 1999
Fifth, revised edition 2002
Reprinted with corrections 2005
Reprinted 2010, 2014

© 2014 International Union of Crystallography

by John Wiley & Sons, Ltd

Registered office: John Wiley & Sons Ltd, The Atrium, Southern Gate, Chichester, West
Sussex, PO19 8SQ, United Kingdom

For details of our global editorial offices, for customer services and for information about
how to apply for permission to reuse the copyright material in this book please see our
website at www.wiley.com.

Technical Editors: D. W. Penfold, M. H. Dacombe, S. E. Barnes and N. J. Ashcroft

Printed in The Netherlands by Printforce Nederland b.v./Alphen aan den Rijn

Contributing authors

H. ARNOLD: Institut für Kristallographie, Rheinisch-Westfälische Technische Hochschule, D-52056 Aachen, Germany (present address: Am Beulardstein 22, D-52072 Aachen, Germany). [2, 5, 11]

M. I. AROYO: Faculty of Physics, University of Sofia, bulv. J. Boucher 5, 1164 Sofia, Bulgaria (present address: Departamento de Fisica de la Materia Condensada, Facultad de Ciencias, Universidad del Pais Vasco, Apartado 644, 48080 Bilbao, Spain). [Computer production of space-group tables]

E. F. BERTAUT†: Laboratoire de Cristallographie, CNRS, Grenoble, France. [4, 13]

Y. BILLIET: Département de Chimie, Faculté des Sciences et Techniques, Université de Bretagne Occidentale, Brest, France (present address: 8 place de Jonquilles, F-29860 Bourg-Blanc, France). [13]

M. J. BUERGER†: Department of Earth and Planetary Sciences, Massachusetts Institute of Technology, Cambridge, MA, USA. [2, 3]

H. BURZLAFF: Universität Erlangen–Nürnberg, Robert-Koch-Strasse 4a, D-91080 Uttenreuth, Germany. [9.1, 12]

J. D. H. DONNAY†: Department of Geological Sciences, McGill University, Montreal, Quebec, Canada. [2]

W. FISCHER: Institut für Mineralogie, Petrologie und Kristallographie, Philipps-Universität, D-35032 Marburg, Germany. [2, 11, 14, 15]

D. S. FOKKEMA: Rekencentrum der Rijksuniversiteit, Groningen, The Netherlands. [Computer production of space-group tables]

B. GRUBER: Department of Applied Mathematics, Faculty of Mathematics and Physics, Charles University, Malostranské nám. 25, CZ-11800 Prague 1, Czech Republic (present address: Sochařská 14, CZ-17000 Prague 7, Czech Republic). [9.3]

TH. HAHN: Institut für Kristallographie, Rheinisch-Westfälische Technische Hochschule, D-52056 Aachen, Germany. [1, 2, 10]

H. KLAPPER: Institut für Kristallographie, Rheinisch-Westfälische Technische Hochschule, D-52056 Aachen, Germany (present address: Mineralogisch-Petrologisches Institut, Universität Bonn, D-53115 Bonn, Germany). [10]

E. KOCH: Institut für Mineralogie, Petrologie und Kristallographie, Philipps-Universität, D-35032 Marburg, Germany. [11, 14, 15]

P. B. KONSTANTINOV: Institute for Nuclear Research and Nuclear Energy, 72 Tzarigradsko Chaussee, BG-1784 Sofia, Bulgaria. [Computer production of space-group tables]

G. A. LANGLET†: Département de Physico-Chimie, CEA, CEN Saclay, Gif sur Yvette, France. [2]

A. LOOIJENGA-VOS: Laboratorium voor Chemische Fysica, Rijksuniversiteit Groningen, The Netherlands (present address: Roland Holstlaan 908, 2624 JK Delft, The Netherlands). [2, 3]

U. MÜLLER: Fachbereich Chemie, Philipps-Universität, D-35032 Marburg, Germany. [15.1, 15.2]

P. M. DE WOLFF†: Laboratorium voor Technische Natuurkunde, Technische Hogeschool, Delft, The Netherlands. [2, 9.2]

H. WONDRATSCHEK: Institut für Kristallographie, Universität, D-76128 Karlsruhe, Germany. [2, 8]

H. ZIMMERMANN: Institut für Angewandte Physik, Lehrstuhl für Kristallographie und Strukturphysik, Universität Erlangen–Nürnberg, Bismarckstrasse 10, D-91054 Erlangen, Germany. [9.1, 12]

† Deceased.

Contents

CONTENTS

Preface to the Fifth, Revised Edition

By Th. Hahn

Volume A of *International Tables for Crystallography* was first published in 1983. Shortly after, in 1985, the *Brief Teaching Edition of Volume A* was prepared, of which the present volume is the Fifth Edition. It is based on the corrected reprint of the Fifth, Revised Edition of Volume A (2005).

The *Teaching Edition* consists of:

(i) complete descriptions of the 17 plane groups, so useful for the teaching of symmetry;

(ii) 24 selected space-group examples, of varying complexity and distributed over all seven crystal systems;

(iii) those basic text sections of Volume A which are necessary for the understanding and handling of space groups (Parts 1, 2, 3 and 5).

Note that space group No. 64 (*Cmce*) provides an example containing the 'double' glide plane *e*.

The purpose of the *Teaching Edition* is threefold:

(i) It should provide a handy (and inexpensive) tool for researchers and students to familiarize themselves with the use of the space-group tables in Volume A.

(ii) It is designed for use in classroom teaching, and with this aim in mind the price has been kept as low as possible. In order to achieve this, the material has been reprinted from Volume A without any changes, except for pagination; hence, this *Teaching Edition* contains references to sections which are only found in Volume A.

(iii) It may serve as a laboratory handbook because the 24 examples include most of the frequently occurring space groups, for both organic and inorganic crystals.

In addition to the 24 space groups given explicitly, further space groups may easily be derived by making use of the general-position entries for the maximal subgroups of types **I** (*translationengleich*) and **IIa** (*klassengleich decentred*) as described in Section 2.2.15.1: The numbers given refer to those coordinate triplets of the general position of the group which are retained in the maximal subgroup and thus characterize the subgroup completely. For those maximal subgroups which conventionally are referred to the same basis vectors and the same origin as the group, the 'standard description', as given in Volume A, is obtained.

This procedure is illustrated by the following example:

For space group No. 199, $I2_13$ (p. 147), the following entries are given under

Maximal non-isomorphic subgroups

I   [3] $I2_11$ ($I2_12_12_1$, 24)   (1; 2; 3; 4)+

which has to be read as

$$(0,0,0)+ \quad (\tfrac{1}{2},\tfrac{1}{2},\tfrac{1}{2})+$$
$$(1)\ x,y,z \qquad (2)\ \bar{x}+\tfrac{1}{2},\bar{y},z+\tfrac{1}{2}$$
$$(3)\ \bar{x},y+\tfrac{1}{2},\bar{z}+\tfrac{1}{2} \quad (4)\ x+\tfrac{1}{2},\bar{y}+\tfrac{1}{2},\bar{z}.$$

This is identical with the general position of space group No. 24, $I2_12_12_1$ (p. 217 of Volume A), which is a maximal *translationengleiche* subgroup of $I2_13$ of index [3].

IIa   [2] $P2_13$ (198)   1; 2; 3; 4; 5; 6; 7; 8; 9; 10; 11; 12

which has to be read as

$$(1)\ x,y,z \quad (2)\ \bar{x}+\tfrac{1}{2},\bar{y},z+\tfrac{1}{2} \quad \cdots \quad (12)\ \bar{y}+\tfrac{1}{2},\bar{z},x+\tfrac{1}{2}.$$

This is identical with the general position of space group No. 198, $P2_13$ (p. 610 of Volume A), which is a maximal *klassengleiche* (decentred) subgroup of $I2_13$ of index [2].

(The other entries under **I** on p. 147 refer to four conjugate maximal *translationengleiche* subgroups of type $R3$ and index [4]; these entries, however, are *not* based on the standard axes and origin of $R3$.)

Similar relations hold for the following examples:

$P\bar{1}$ (2)	yields	$P1$ (1)
$C12/m1$ (12)	yields	$C121$ (5); $C1m1$ (8); $P12/m1$ (10)
$C12/c1$ (15)	yields	$C1c1$ (9); $P12/c1$ (13); $P12_1/n1$ (14)
$Pmna$ (53)	yields	$P112_1/a$ (14); $P12/n1$ (13); $Pmn2_1$ (31)
$Cmce$ (*Cmca*) (64)	yields	$Pbca$ (61)
$R\bar{3}m$ (166)	yields	$R32$ (155); $R\bar{3}$ (148); $R3m$ (160); $P\bar{3}m1$ (164)
$P6_3/mmc$ (194)	yields	$P6_322$ (182); $P6_3/m$ (176); $P6_3mc$ (186); $P\bar{3}m1$ (164); $P\bar{3}1c$ (163); $P\bar{6}2c$ (190)
$I2_13$ (199)	yields	$I2_12_12_1$ (24); $P2_13$ (198)
$Fm\bar{3}m$ (225)	yields	$Fm\bar{3}$ (202); $F432$ (209); $F\bar{4}3m$ (216); $Pm\bar{3}m$ (221); $Pn\bar{3}m$ (224)
$Fd\bar{3}m$ (227, origin 1)	yields	$Fd\bar{3}$ (203); $F4_132$ (210); $F\bar{4}3m$ (216).

It is an interesting exercise to complete this list for the 24 selected space groups and to extend it even to those maximal subgroups where the origin, the basis vectors, or both, are different from the group; in fact, to encourage this kind of 'playing' with space groups is one of the intentions of the *Teaching Edition*.

The Editor wishes to extend his sincere thanks to the International Union of Crystallography for making this *Teaching Edition* possible, to D. W. Penfold, M. H. Dacombe, S. E. Barnes and N. J. Ashcroft (Chester) for its technical preparation, and to a number of colleagues for counsel on the selection of material, especially D. W. J. Cruickshank (Manchester) and H. Wondratschek (Karlsruhe).

Aachen, November 2005   Theo Hahn

1. SYMBOLS AND TERMS USED IN THIS VOLUME

By Th. Hahn

1.1. Printed symbols for crystallographic items

By Th. Hahn

1.1.1. Vectors, coefficients and coordinates

Printed symbol	Explanation
$\mathbf{a}, \mathbf{b}, \mathbf{c}$; or $\mathbf{a}_i$	Basis vectors of the direct lattice
a, b, c	Lengths of basis vectors, lengths of cell edges ⎫ Lattice or cell parameters
α, β, γ	Interaxial (lattice) angles $\mathbf{b} \wedge \mathbf{c}$, $\mathbf{c} \wedge \mathbf{a}$, $\mathbf{a} \wedge \mathbf{b}$ ⎭
V	Cell volume of the direct lattice
$\mathbf{G}$	Matrix of the geometrical coefficients (metric tensor) of the direct lattice
g_{ij}	Element of metric matrix (tensor) $\mathbf{G}$
$\mathbf{r}$; or $\mathbf{x}$	Position vector (of a point or an atom)
r	Length of the position vector $\mathbf{r}$
$x\mathbf{a}, y\mathbf{b}, z\mathbf{c}$	Components of the position vector $\mathbf{r}$
x, y, z; or x_i	Coordinates of a point (location of an atom) expressed in units of a, b, c; coordinates of end point of position vector $\mathbf{r}$; coefficients of position vector $\mathbf{r}$
$\boldsymbol{x} = \begin{pmatrix} x \\ y \\ z \end{pmatrix} = \begin{pmatrix} x_1 \\ x_2 \\ x_3 \end{pmatrix}$	Column of point coordinates or vector coefficients
$\mathbf{t}$	Translation vector
t	Length of the translation vector $\mathbf{t}$
t_1, t_2, t_3; or t_i	Coefficients of translation vector $\mathbf{t}$
$\boldsymbol{t} = \begin{pmatrix} t_1 \\ t_2 \\ t_3 \end{pmatrix}$	Column of coefficients of translation vector $\mathbf{t}$
$\mathbf{u}$	Vector with integral coefficients
u, v, w; or u_i	Integers, coordinates of a (primitive) lattice point; coefficients of vector $\mathbf{u}$
$\boldsymbol{u} = \begin{pmatrix} u \\ v \\ w \end{pmatrix} = \begin{pmatrix} u_1 \\ u_2 \\ u_3 \end{pmatrix}$	Column of integral point coordinates or vector coefficients
$\mathbf{o}$	Zero vector
$\boldsymbol{o}$	Column of zero coefficients
$\mathbf{a}', \mathbf{b}', \mathbf{c}'$; or $\mathbf{a}'_i$	New basis vectors after a transformation of the coordinate system (basis transformation)
$\mathbf{r}'$; or $\mathbf{x}'$; x', y', z'; or x'_i	Position vector and point coordinates after a transformation of the coordinate system (basis transformation)
$\tilde{\mathbf{r}}$; or $\tilde{\mathbf{x}}$; $\tilde{x}, \tilde{y}, \tilde{z}$; or $\tilde{x}_i$	New position vector and point coordinates after a symmetry operation (motion)

1.1.2. Directions and planes

Printed symbol	Explanation
$[uvw]$	Indices of a lattice direction (zone axis)
$\langle uvw \rangle$	Indices of a set of all symmetrically equivalent lattice directions
(hkl)	Indices of a crystal face, or of a single net plane (Miller indices)
$(hkil)$	Indices of a crystal face, or of a single net plane, for the hexagonal axes $\mathbf{a}_1$, $\mathbf{a}_2$, $\mathbf{a}_3$, $\mathbf{c}$ (Bravais–Miller indices)
$\{hkl\}$	Indices of a set of all symmetrically equivalent crystal faces ('crystal form'), or net planes
$\{hkil\}$	Indices of a set of all symmetrically equivalent crystal faces ('crystal form'), or net planes, for the hexagonal axes $\mathbf{a}_1, \mathbf{a}_2, \mathbf{a}_3, \mathbf{c}$
hkl	Indices of the Bragg reflection (Laue indices) from the set of parallel equidistant net planes (hkl)
d_{hkl}	Interplanar distance, or spacing, of neighbouring net planes (hkl)

1.1.3. Reciprocal space

Printed symbol	Explanation
$\mathbf{a}^*, \mathbf{b}^*, \mathbf{c}^*$; or $\mathbf{a}^*_i$	Basis vectors of the reciprocal lattice
a^*, b^*, c^*	Lengths of basis vectors of the reciprocal lattice
$\alpha^*, \beta^*, \gamma^*$	Interaxial (lattice) angles of the reciprocal lattice $\mathbf{b}^* \wedge \mathbf{c}^*$, $\mathbf{c}^* \wedge \mathbf{a}^*$, $\mathbf{a}^* \wedge \mathbf{b}^*$
$\mathbf{r}^*$; or $\mathbf{h}$	Reciprocal-lattice vector
h, k, l; or h_i	Coordinates of a reciprocal-lattice point, expressed in units of a^*, b^*, c^*, coefficients of the reciprocal-lattice vector $\mathbf{r}^*$
V^*	Cell volume of the reciprocal lattice
$\mathbf{G}^*$	Matrix of the geometrical coefficients (metric tensor) of the reciprocal lattice

1.1.4. Functions

Printed symbol	Explanation
$\rho(xyz)$	Electron density at the point x, y, z
$P(xyz)$	Patterson function at the point x, y, z
$F(hkl)$; or F	Structure factor (of the unit cell), corresponding to the Bragg reflection hkl
$\lvert F(hkl) \rvert$; or $\lvert F \rvert$	Modulus of the structure factor $F(hkl)$
$\alpha(hkl)$; or α	Phase angle of the structure factor $F(hkl)$

1.1.5. Spaces

Printed symbol	Explanation
n	Dimension of a space
X	Point
$\tilde{X}$	Image of a point X after a symmetry operation (motion)
E^n	(Euclidean) point space of dimension n
$\mathbf{V}^n$	Vector space of dimension n
$\mathbf{L}$	Vector lattice
L	Point lattice

1.1.6. Motions and matrices

Printed symbol	Explanation
W; M	Symmetry operation; motion
$(\boldsymbol{W}, \boldsymbol{w})$	Symmetry operation W, described by an $(n \times n)$ matrix $\boldsymbol{W}$ and an $(n \times 1)$ column $\boldsymbol{w}$
$\mathbb{W}$	Symmetry operation W, described by an $(n + 1) \times (n + 1)$ 'augmented' matrix
$\boldsymbol{I}$	$(n \times n)$ unit matrix
T	Translation
$(\boldsymbol{I}, \boldsymbol{t})$	Translation T, described by the $(n \times n)$ unit matrix $\boldsymbol{I}$ and an $(n \times 1)$ column $\boldsymbol{t}$
$\mathbb{T}$	Translation T, described by an $(n + 1) \times (n + 1)$ 'augmented' matrix
I	Identity operation
$(\boldsymbol{I}, \boldsymbol{o})$	Identity operation I, described by the $(n \times n)$ unit matrix $\boldsymbol{I}$ and the $(n \times 1)$ column $\boldsymbol{o}$
$\mathbb{I}$	Identity operation I, described by the $(n + 1) \times (n + 1)$ 'augmented' unit matrix

Printed symbol	Explanation
$\mathbb{r}$, or $\mathbb{x}$	Position vector (of a point or an atom), described by an $(n + 1) \times 1$ 'augmented' column
$(\boldsymbol{P}, \boldsymbol{p})$; or $(\boldsymbol{S}, \boldsymbol{s})$	Transformation of the coordinate system, described by an $(n \times n)$ matrix $\boldsymbol{P}$ or $\boldsymbol{S}$ and an $(n \times 1)$ column $\boldsymbol{p}$ or $\boldsymbol{s}$
$\mathbb{P}$; or $\mathbb{S}$	Transformation of the coordinate system, described by an $(n + 1) \times (n + 1)$ 'augmented' matrix
$(\boldsymbol{Q}, \boldsymbol{q})$	Inverse transformation of $(\boldsymbol{P}, \boldsymbol{p})$
$\mathbb{Q}$	Inverse transformation of $\mathbb{P}$

1.1.7. Groups

Printed symbol	Explanation
$\mathcal{G}$	Space group
$\mathcal{T}$	Group of all translations of $\mathcal{G}$
$\mathcal{S}$	Supergroup; also used for site-symmetry group
$\mathcal{H}$	Subgroup
$\mathcal{E}$	Group of all motions (Euclidean group)
$\mathcal{A}$	Group of all affine mappings (affine group)
$\mathcal{N}_{\mathcal{E}}(\mathcal{G})$; or $\mathcal{N}_{\mathcal{A}}(\mathcal{G})$	Euclidean or affine normalizer of a space group $\mathcal{G}$
$\mathcal{P}$	Point group
$\mathcal{C}$	*Eigensymmetry* (inherent symmetry) group
$[i]$	Index i of sub- or supergroup
G	Element of a space group $\mathcal{G}$

1.2. Printed symbols for conventional centring types

By Th. Hahn

1.2.1. Printed symbols for the conventional centring types of one-, two- and three-dimensional cells

For 'reflection conditions', see Tables 2.2.13.1 and 2.2.13.3. For the new centring symbol S, see Note (iii) below.

Printed symbol	Centring type of cell	Number of lattice points per cell	Coordinates of lattice points within cell
One dimension			
p	Primitive	1	0
Two dimensions			
p	Primitive	1	0, 0
c	Centred	2	$0, 0; \frac{1}{2}, \frac{1}{2}$
h^*	Hexagonally centred	3	$0, 0; \frac{2}{3}, \frac{1}{3}; \frac{1}{3}, \frac{2}{3}$
Three dimensions			
P	Primitive	1	0, 0, 0
C	C-face centred	2	$0, 0, 0; \frac{1}{2}, \frac{1}{2}, 0$
A	A-face centred	2	$0, 0, 0; 0, \frac{1}{2}, \frac{1}{2}$
B	B-face centred	2	$0, 0, 0; \frac{1}{2}, 0, \frac{1}{2}$
I	Body centred	2	$0, 0, 0; \frac{1}{2}, \frac{1}{2}, \frac{1}{2}$
F	All-face centred	4	$0, 0, 0; \frac{1}{2}, \frac{1}{2}, 0; 0, \frac{1}{2}, \frac{1}{2}; \frac{1}{2}, 0, \frac{1}{2}$
R†	Rhombohedrally centred (description with 'hexagonal axes')	3	$\left\{\begin{array}{l} 0, 0, 0; \frac{2}{3}, \frac{1}{3}, \frac{1}{3}; \frac{1}{3}, \frac{2}{3}, \frac{2}{3} \text{ ('obverse setting')} \\ 0, 0, 0; \frac{1}{3}, \frac{2}{3}, \frac{1}{3}; \frac{2}{3}, \frac{1}{3}, \frac{2}{3} \text{ ('reverse setting')} \end{array}\right.$
	Primitive (description with 'rhombohedral axes')	1	0, 0, 0
H‡	Hexagonally centred	3	$0, 0, 0; \frac{2}{3}, \frac{1}{3}, 0; \frac{1}{3}, \frac{2}{3}, 0$

* The two-dimensional triple hexagonal cell h is an alternative description of the hexagonal plane net, as illustrated in Fig. 5.1.3.8. It is not used for systematic plane-group description in this volume; it is introduced, however, in the sub- and supergroup entries of the plane-group tables (Part 6). Plane-group symbols for the h cell are listed in Chapter 4.2. Transformation matrices are contained in Table 5.1.3.1.

† In the space-group tables (Part 7), as well as in *IT* (1935) and *IT* (1952) [for reference notation, see footnote on first page of Chapter 2.1], the seven rhombohedral R space groups are presented with two descriptions, one based on *hexagonal axes* (triple cell), one on *rhombohedral axes* (primitive cell). In the present volume, as well as in *IT* (1952), the *obverse* setting of the triple hexagonal cell R is used. Note that in *IT* (1935) the *reverse* setting was employed. The two settings are related by a rotation of the hexagonal cell with respect to the rhombohedral lattice around a threefold axis, involving a rotation angle of 60°, 180° or 300° (*cf.* Fig. 5.1.3.6). Further details may be found in Chapter 2.1, Section 4.3.5 and Chapter 9.1. Transformation matrices are contained in Table 5.1.3.1.

‡ The triple hexagonal cell H is an alternative description of the hexagonal Bravais lattice, as illustrated in Fig. 5.1.3.8. It was used for systematic space-group description in *IT* (1935), but replaced by P in *IT* (1952). In the space-group tables of this volume (Part 7), it is only used in the sub- and supergroup entries (*cf.* Section 2.2.15). Space-group symbols for the H cell are listed in Section 4.3.5. Transformation matrices are contained in Table 5.1.3.1.

1.2.2. Notes on centred cells

(i) The centring type of a cell may change with a change of the basis vectors; in particular, a primitive cell may become a centred cell and *vice versa*. Examples of relevant transformation matrices are contained in Table 5.1.3.1.

(ii) Section 1.2.1 contains only those conventional centring symbols which occur in the Hermann–Mauguin space-group symbols. There exist, of course, further kinds of centred cells which are unconventional; an interesting example is provided by the triple rhombohedral D cell, described in Section 4.3.5.3.

(iii) For the use of the letter S as a new general, setting-independent 'centring symbol' for monoclinic and orthorhombic Bravais lattices see Chapter 2.1, especially Table 2.1.2.1, and de Wolff *et al.* (1985).

(iv) Symbols for crystal families and Bravais lattices in one, two and three dimensions are listed in Table 2.1.2.1 and are explained in the *Nomenclature Report* by de Wolff *et al.* (1985).

1.3. Printed symbols for symmetry elements

By Th. Hahn

1.3.1. Printed symbols for symmetry elements and for the corresponding symmetry operations in one, two and three dimensions

For 'reflection conditions', see Tables 2.2.13.2 and 2.2.13.3.

Printed symbol	Symmetry element and its orientation	Defining symmetry operation with glide or screw vector
m	Reflection plane, mirror plane	Reflection through the plane
	Reflection line, mirror line (two dimensions)	Reflection through the line
	Reflection point, mirror point (one dimension)	Reflection through the point
a, b or c	'Axial' glide plane	Glide reflection through the plane, with glide vector
a	$\perp [010]$ or $\perp [001]$	$\frac{1}{2}\mathbf{a}$
b	$\perp [001]$ or $\perp [100]$	$\frac{1}{2}\mathbf{b}$
c †	$\perp [100]$ or $\perp [010]$	$\frac{1}{2}\mathbf{c}$
	$\perp [1\bar{1}0]$ or $\perp [110]$	$\frac{1}{2}\mathbf{c}$
	$\perp [100]$ or $\perp [010]$ or $\perp [\bar{1}10]$	$\frac{1}{2}\mathbf{c}$ ⎫ hexagonal coordinate system
	$\perp [1\bar{1}0]$ or $\perp [120]$ or $\perp [\bar{2}\bar{1}0]$	$\frac{1}{2}\mathbf{c}$ ⎭
e ‡	'Double' glide plane (in centred cells only)	*Two* glide reflections through *one* plane, with perpendicular glide vectors
	$\perp [001]$	$\frac{1}{2}\mathbf{a}$ *and* $\frac{1}{2}\mathbf{b}$
	$\perp [100]$	$\frac{1}{2}\mathbf{b}$ *and* $\frac{1}{2}\mathbf{c}$
	$\perp [010]$	$\frac{1}{2}\mathbf{a}$ *and* $\frac{1}{2}\mathbf{c}$
	$\perp [1\bar{1}0]; \perp [110]$	$\frac{1}{2}(\mathbf{a}+\mathbf{b})$ *and* $\frac{1}{2}\mathbf{c}; \frac{1}{2}(\mathbf{a}-\mathbf{b})$ *and* $\frac{1}{2}\mathbf{c}$
	$\perp [01\bar{1}]; \perp [011]$	$\frac{1}{2}(\mathbf{b}+\mathbf{c})$ *and* $\frac{1}{2}\mathbf{a}; \frac{1}{2}(\mathbf{b}-\mathbf{c})$ *and* $\frac{1}{2}\mathbf{a}$
	$\perp [\bar{1}01]; \perp [101]$	$\frac{1}{2}(\mathbf{a}+\mathbf{c})$ *and* $\frac{1}{2}\mathbf{b}; \frac{1}{2}(\mathbf{a}-\mathbf{c})$ *and* $\frac{1}{2}\mathbf{b}$
n	'Diagonal' glide plane	Glide reflection through the plane, with glide vector
	$\perp [001]; \perp [100]; \perp [010]$	$\frac{1}{2}(\mathbf{a}+\mathbf{b}); \frac{1}{2}(\mathbf{b}+\mathbf{c}); \frac{1}{2}(\mathbf{a}+\mathbf{c})$
	$\perp [1\bar{1}0]$ or $\perp [01\bar{1}]$ or $\perp [\bar{1}01]$	$\frac{1}{2}(\mathbf{a}+\mathbf{b}+\mathbf{c})$
	$\perp [110]; \perp [011]; \perp [101]$	$\frac{1}{2}(-\mathbf{a}+\mathbf{b}+\mathbf{c}); \frac{1}{2}(\mathbf{a}-\mathbf{b}+\mathbf{c}); \frac{1}{2}(\mathbf{a}+\mathbf{b}-\mathbf{c})$
d §	'Diamond' glide plane	Glide reflection through the plane, with glide vector
	$\perp [001]; \perp [100]; \perp [010]$	$\frac{1}{4}(\mathbf{a}\pm\mathbf{b}); \frac{1}{4}(\mathbf{b}\pm\mathbf{c}); \frac{1}{4}(\pm\mathbf{a}+\mathbf{c})$
	$\perp [1\bar{1}0]; \perp [01\bar{1}]; \perp [\bar{1}01]$	$\frac{1}{4}(\mathbf{a}+\mathbf{b}\pm\mathbf{c}); \frac{1}{4}(\pm\mathbf{a}+\mathbf{b}+\mathbf{c}); \frac{1}{4}(\mathbf{a}\pm\mathbf{b}+\mathbf{c})$
	$\perp [110]; \perp [011]; \perp [101]$	$\frac{1}{4}(-\mathbf{a}+\mathbf{b}\pm\mathbf{c}); \frac{1}{4}(\pm\mathbf{a}-\mathbf{b}+\mathbf{c}); \frac{1}{4}(\mathbf{a}\pm\mathbf{b}-\mathbf{c})$
g	Glide line (two dimensions)	Glide reflection through the line, with glide vector
	$\perp [01]; \perp [10]$	$\frac{1}{2}\mathbf{a}; \frac{1}{2}\mathbf{b}$
1	None	Identity
2, 3, 4, 6	n-fold rotation axis, n	Counter-clockwise rotation of $360/n$ degrees around the axis (see Note viii)
	n-fold rotation point, n (two dimensions)	Counter-clockwise rotation of $360/n$ degrees around the point
$\bar{1}$	Centre of symmetry, inversion centre	Inversion through the point
$\bar{2} = m,$ ¶ $\bar{3}, \bar{4}, \bar{6}$	Rotoinversion axis, $\bar{n}$, and inversion point on the axis††	Counter-clockwise rotation of $360/n$ degrees around the axis, followed by inversion through the point on the axis†† (see Note viii)
2_1 $3_1, 3_2$ $4_1, 4_2, 4_3$ $6_1, 6_2, 6_3, 6_4, 6_5$	n-fold screw axis, n_p	Right-handed screw rotation of $360/n$ degrees around the axis, with screw vector (pitch) (p/n) $\mathbf{t}$; here $\mathbf{t}$ is the shortest lattice translation vector parallel to the axis in the direction of the screw

† In the rhombohedral space-group symbols $R3c$ (161) and $R\bar{3}c$ (167), the symbol c refers to the description with 'hexagonal axes'; *i.e.* the glide vector is $\frac{1}{2}\mathbf{c}$, along [001]. In the description with 'rhombohedral axes', this glide vector is $\frac{1}{2}(\mathbf{a}+\mathbf{b}+\mathbf{c})$, along [111], *i.e.* the symbol of the glide plane would be n: *cf.* Section 4.3.5.

‡ For further explanations of the 'double' glide plane e, see Note (x) below.

§ Glide planes d occur only in orthorhombic F space groups, in tetragonal I space groups, and in cubic I and F space groups. They always occur in pairs with alternating glide vectors, for instance $\frac{1}{4}(\mathbf{a}+\mathbf{b})$ and $\frac{1}{4}(\mathbf{a}-\mathbf{b})$. The second power of a glide reflection d is a centring vector.

¶ Only the symbol m is used in the Hermann–Mauguin symbols, for both point groups and space groups.

†† The inversion point is a centre of symmetry if n is odd.

1.3.2. Notes on symmetry elements and symmetry operations

(i) Section 1.3.1 contains only those symmetry elements and symmetry operations which occur in the Hermann–Mauguin symbols of point groups and space groups. Further so-called 'additional symmetry elements' are described in Chapter 4.1 and listed in Tables 4.2.1.1 and 4.3.2.1 in the form of 'extended Hermann–Mauguin symbols'.

(ii) The printed symbols of symmetry elements (symmetry operations), except for glide planes (glide reflections), are independent of the choice and the labelling of the basis vectors and of the origin. The symbols of glide planes (glide reflections), however, may change with a change of the basis vectors. For this reason, the possible orientations of glide planes and the glide vectors of the corresponding operations are listed explicitly in columns 2 and 3.

(iii) In space groups, further kinds of glide planes and glide reflections (called g) occur which are not used in the Hermann–Mauguin symbols. They are listed in the space-group tables (Part 7) under *Symmetry operations* and in Table 4.3.2.1 for the tetragonal and cubic space groups; they are explained in Sections 2.2.9 and 11.1.2.

(iv) Whereas the term 'symmetry operation' is well defined (*cf.* Section 8.1.3), the word 'symmetry element' is used by crystallographers in a variety of often rather loose meanings. In 1989, the International Union of Crystallography published a *Nomenclature Report* which first defines a 'geometric element' as a geometric item that allows the fixed points of a symmetry operation (after removal of any intrinsic glide or screw translation) to be located and oriented in a coordinate system. A 'symmetry element' then is defined as a concept with a double meaning, namely the combination of a geometric element with the set of symmetry operations having this geometric element in common ('element set'). For further details and tables, see de Wolff *et al.* (1989) and Flack *et al.* (2000).

(v) To each glide plane, infinitely many different glide reflections belong, because to each glide vector listed in column 3 any lattice translation vector parallel to the glide plane may be added; this includes centring vectors of centred cells. Each resulting vector is a glide vector of a new glide reflection but with the same plane as the geometric element. Any of these glide operations can be used as a 'defining operation'.

Examples
(1) Glide plane $n \perp [001]$: All vectors $(u + \frac{1}{2})\mathbf{a} + (v + \frac{1}{2})\mathbf{b}$ are glide vectors (u, v any integers); this includes $\frac{1}{2}(\mathbf{a} + \mathbf{b})$, $\frac{1}{2}(\mathbf{a} - \mathbf{b})$, $\frac{1}{2}(-\mathbf{a} + \mathbf{b})$, $\frac{1}{2}(-\mathbf{a} - \mathbf{b})$.
(2) Glide plane $e \perp [001]$ in a C-centred cell: All vectors $(u + \frac{1}{2})\mathbf{a} + v\mathbf{b}$ and $u\mathbf{a} + (v + \frac{1}{2})\mathbf{b}$ are glide vectors, this includes $\frac{1}{2}\mathbf{a}$ and $\frac{1}{2}\mathbf{b}$ (which are related by the centring vector), *i.e.* the glide plane e is at the same time a glide plane a *and* a glide plane b; for this 'double' glide plane e see Note (x) below.
(3) Glide plane $c \perp [1\bar{1}0]$ in an F-centred cell: All vectors $\frac{1}{2}u(\mathbf{a} + \mathbf{b}) + (v + \frac{1}{2})\mathbf{c}$ are glide vectors; this includes $\frac{1}{2}\mathbf{c}$ and $\frac{1}{2}(\mathbf{a} + \mathbf{b} + \mathbf{c})$, *i.e.* the glide plane c is at the same time a glide plane n.

(vi) If among the infinitely many glide operations of the element set of a symmetry plane there exists *one* operation with glide vector zero, then this symmetry element is a mirror plane.

(vii) Similar considerations apply to screw axes; to the screw vector defined in column 3 any lattice translation vector parallel to the screw axis may be added. Again, this includes centring vectors of centred cells.

Example
Screw axis $3_1 \parallel [111]$ in a cubic primitive cell. For the first power (right-handed screw rotation of 120°), all vectors $(u + \frac{1}{3})(\mathbf{a} + \mathbf{b} + \mathbf{c})$ are screw vectors; this includes $\frac{1}{3}(\mathbf{a} + \mathbf{b} + \mathbf{c})$, $\frac{4}{3}(\mathbf{a} + \mathbf{b} + \mathbf{c})$, $-\frac{2}{3}(\mathbf{a} + \mathbf{b} + \mathbf{c})$. For the second power (right-handed screw rotation of 240°), all vectors $(u + \frac{2}{3})(\mathbf{a} + \mathbf{b} + \mathbf{c})$ are screw vectors; this includes $\frac{2}{3}(\mathbf{a} + \mathbf{b} + \mathbf{c})$, $\frac{5}{3}(\mathbf{a} + \mathbf{b} + \mathbf{c})$; $-\frac{1}{3}(\mathbf{a} + \mathbf{b} + \mathbf{c})$. The third power corresponds to all lattice vectors $u(\mathbf{a} + \mathbf{b} + \mathbf{c})$.

Again, if *one* of the screw vectors is zero, the symmetry element is a rotation axis.

(viii) In the space-group tables, under *Symmetry operations*, for rotations, screw rotations and roto-inversions, the 'sense of rotation' is indicated by symbols like 3^+, $\bar{4}^-$ *etc.*; this is explained in Section 11.1.2.

(ix) The members of the following pairs of screw axes are 'enantiomorphic', *i.e.* they can be considered as a right- and a left-handed screw, respectively, with the same screw vector: 3_1, 3_2; 4_1, 4_3; 6_1, 6_5; 6_2, 6_4. The following screw axes are 'neutral', *i.e.* they contain left- *and* right-handed screws with the same screw vector: 2_1; 4_2; 6_3.

(x) In the third *Nomenclature Report* of the IUCr (de Wolff *et al.*, 1992), two new printed symbols for glide planes were proposed: e for 'double' glide planes and k for 'transverse' glide planes.

For the e glide planes, new graphical symbols were introduced (*cf.* Sections 1.4.1, 1.4.2, 1.4.3 and Note iv in 1.4.4); they are applied to the diagrams of the relevant space groups: Seven orthorhombic A-, C- and F-space groups, five tetragonal I-space groups, and five cubic F- and I-space groups. The e glide plane occurs only in centred cells and is defined by *one* plane with *two* perpendicular glide vectors related by a centring translation; thus, in *Cmma* (67), two glide operations a and b through the plane $xy0$ occur, their glide vectors being related by the centring vector $\frac{1}{2}(\mathbf{a} + \mathbf{b})$; the symbol e removes the ambiguity between the symbols a and b.

For five space groups, the Hermann–Mauguin symbol has been modified:

Space group No.	39	41	64	67	68
New symbol:	*Aem*2	*Aea*2	*Cmce*	*Cmme*	*Ccce*
Former symbol:	*Abm*2	*Aba*2	*Cmca*	*Cmma*	*Ccca*

The new symbol is now the standard one; it is indicated in the headline of these space groups, while the former symbol is given underneath.

For the k glide planes, no new graphical symbol and no modification of a space-group symbol are proposed.

1.4. Graphical symbols for symmetry elements in one, two and three dimensions

By Th. Hahn

1.4.1. Symmetry planes normal to the plane of projection (three dimensions) and symmetry lines in the plane of the figure (two dimensions)

Symmetry plane or symmetry line	Graphical symbol	Glide vector in units of lattice translation vectors parallel and normal to the projection plane	Printed symbol
Reflection plane, mirror plane Reflection line, mirror line (two dimensions)	——————	None	m
'Axial' glide plane Glide line (two dimensions)	− − − − −	$\frac{1}{2}$ lattice vector along line in projection plane $\frac{1}{2}$ lattice vector along line in figure plane	a, b or c g
'Axial' glide plane	··············	$\frac{1}{2}$ lattice vector normal to projection plane	a, b or c
'Double' glide plane* (in centred cells only)	··−··−·−··	*Two* glide vectors: $\frac{1}{2}$ along line parallel to projection plane and $\frac{1}{2}$ normal to projection plane	e
'Diagonal' glide plane	−·−·−·−	*One* glide vector with *two* components: $\frac{1}{2}$ along line parallel to projection plane, $\frac{1}{2}$ normal to projection plane	n
'Diamond' glide plane† (pair of planes; in centred cells only)	−·−◄·− −·−►·−	$\frac{1}{4}$ along line parallel to projection plane, combined with $\frac{1}{4}$ normal to projection plane (arrow indicates direction parallel to the projection plane for which the normal component is positive)	d

* For further explanations of the 'double' glide plane e see Note (iv) below and Note (x) in Section 1.3.2.
† See footnote § to Section 1.3.1.

1.4.2. Symmetry planes parallel to the plane of projection

Symmetry plane	Graphical symbol*	Glide vector in units of lattice translation vectors parallel to the projection plane	Printed symbol
Reflection plane, mirror plane		None	m
'Axial' glide plane		$\frac{1}{2}$ lattice vector in the direction of the arrow	a, b or c
'Double' glide plane† (in centred cells only)		*Two* glide vectors: $\frac{1}{2}$ in either of the directions of the two arrows	e
'Diagonal' glide plane		*One* glide vector with *two* components $\frac{1}{2}$ in the direction of the arrow	n
'Diamond' glide plane‡ (pair of planes; in centred cells only)		$\frac{1}{2}$ in the direction of the arrow; the glide vector is always half of a centring vector, *i.e.* one quarter of a diagonal of the conventional face-centred cell	d

* The symbols are given at the upper left corner of the space-group diagrams. A fraction h attached to a symbol indicates two symmetry planes with 'heights' h and $h + \frac{1}{2}$ above the plane of projection; *e.g.* $\frac{1}{8}$ stands for $h = \frac{1}{8}$ and $\frac{5}{8}$. No fraction means $h = 0$ and $\frac{1}{2}$ (*cf.* Section 2.2.6).
† For further explanations of the 'double' glide plane e see Note (iv) below and Note (x) in Section 1.3.2.
‡ See footnote § to Section 1.3.1.

1.4.3. Symmetry planes inclined to the plane of projection (in cubic space groups of classes $\bar{4}3m$ and $m\bar{3}m$ only)

Symmetry plane	Graphical symbol* for planes normal to		Glide vector in units of lattice translation vectors for planes normal to		Printed symbol
	[011] and [01$\bar{1}$]	[101] and [10$\bar{1}$]	[011] and [01$\bar{1}$]	[101] and [10$\bar{1}$]	
Reflection plane, mirror plane			None	None	m
'Axial' glide plane			$\frac{1}{2}$ lattice vector along [100]	$\frac{1}{2}$ lattice vector along [010]	a or b
'Axial' glide plane			$\frac{1}{2}$ lattice vector along [01$\bar{1}$] or along [011]	$\frac{1}{2}$ lattice vector along [10$\bar{1}$] or along [101]	
'Double' glide plane† [in space groups $I\bar{4}3m$ (217) and $Im\bar{3}m$ (229) only]			*Two* glide vectors: $\frac{1}{2}$ along [100] *and* $\frac{1}{2}$ along [01$\bar{1}$] or $\frac{1}{2}$ along [011]	*Two* glide vectors: $\frac{1}{2}$ along [010] *and* $\frac{1}{2}$ along [10$\bar{1}$] or $\frac{1}{2}$ along [101]	e
'Diagonal' glide plane			*One* glide vector: $\frac{1}{2}$ along [11$\bar{1}$] or along [111]‡	*One* glide vector: $\frac{1}{2}$ along [11$\bar{1}$] or along [111]‡	n
'Diamond' glide plane¶ (pair of planes; in centred cells only)			$\frac{1}{2}$ along [1$\bar{1}$1] or along [111]§	$\frac{1}{2}$ along [$\bar{1}$11] or along [111]§	d
			$\frac{1}{2}$ along [$\bar{1}\bar{1}$1] or along [111]§	$\frac{1}{2}$ along [$\bar{1}\bar{1}$1] or along [1$\bar{1}$1]§	

* The symbols represent orthographic projections. In the cubic space-group diagrams, complete orthographic projections of the symmetry elements around high-symmetry points, such as $0,0,0; \frac{1}{2},0,0; \frac{1}{4},\frac{1}{4},0$, are given as 'inserts'.

† For further explanations of the 'double' glide plane e see Note (iv) below and Note (x) in Section 1.3.2.

‡ In the space groups $F\bar{4}3m$ (216), $Fm\bar{3}m$ (225) and $Fd\bar{3}m$ (227), the shortest lattice translation vectors in the glide directions are $\mathbf{t}(1,\frac{1}{2},\frac{\bar{1}}{2})$ or $\mathbf{t}(1,\frac{1}{2},\frac{1}{2})$ and $\mathbf{t}(\frac{1}{2},1,\frac{\bar{1}}{2})$ or $\mathbf{t}(\frac{1}{2},1,\frac{1}{2})$, respectively.

§ The glide vector is half of a centring vector, *i.e.* one quarter of the diagonal of the conventional body-centred cell in space groups $I\bar{4}3d$ (220) and $Ia\bar{3}d$ (230).

¶ See footnote § to Section 1.3.1.

1.4.4. Notes on graphical symbols of symmetry planes

(i) The *graphical* symbols and their explanations (columns 2 and 3) are independent of the projection direction and the labelling of the basis vectors. They are, therefore, applicable to any projection diagram of a space group. The *printed* symbols of *glide planes* (column 4), however, may change with a change of the basis vectors, as shown by the following example.

In the rhombohedral space groups $R3c$ (161) and $R\bar{3}c$ (167), the dotted line refers to a c glide when described with 'hexagonal axes' and projected along [001]; for a description with 'rhombohedral axes' and projection along [111], the same dotted glide plane would be called n. The dash-dotted n glide in the hexagonal description becomes an a, b or c glide in the rhombohedral description; *cf.* footnote † to Section 1.3.1.

(ii) The graphical symbols for glide planes in column 2 are not only used for the glide planes defined in Chapter 1.3, but also for the further glide planes g which are mentioned in Section 1.3.2 (Note x) and listed in Table 4.3.2.1; they are explained in Sections 2.2.9 and 11.1.2.

(iii) In monoclinic space groups, the 'parallel' glide vector of a glide plane may be along a lattice translation vector which is inclined to the projection plane.

(iv) In 1992, the International Union of Crystallography introduced the 'double' glide plane e and the graphical symbol ·· –·· – for e glide planes oriented 'normal' and 'inclined' to the plane of projection (de Wolff *et al.*, 1992); for details of e glide planes see Chapter 1.3. Note that the graphical symbol for e glide planes oriented 'parallel' to the projection plane has already been used in *IT* (1935) and *IT* (1952).

1.4.5. Symmetry axes normal to the plane of projection and symmetry points in the plane of the figure

Symmetry axis or symmetry point	Graphical symbol*	Screw vector of a right-handed screw rotation in units of the shortest lattice translation vector parallel to the axis	Printed symbol (partial elements in parentheses)
Identity	None	None	1
Twofold rotation axis / Twofold rotation point (two dimensions)		None	2
Twofold screw axis: '2 sub 1'		$\frac{1}{2}$	2_1
Threefold rotation axis / Threefold rotation point (two dimensions)		None	3
Threefold screw axis: '3 sub 1'		$\frac{1}{3}$	3_1
Threefold screw axis: '3 sub 2'		$\frac{2}{3}$	3_2
Fourfold rotation axis / Fourfold rotation point (two dimensions)		None	4 (2)
Fourfold screw axis: '4 sub 1'		$\frac{1}{4}$	4_1 (2_1)
Fourfold screw axis: '4 sub 2'		$\frac{1}{2}$	4_2 (2)
Fourfold screw axis: '4 sub 3'		$\frac{3}{4}$	4_3 (2_1)
Sixfold rotation axis / Sixfold rotation point (two dimensions)		None	6 (3,2)
Sixfold screw axis: '6 sub 1'		$\frac{1}{6}$	6_1 $(3_1, 2_1)$
Sixfold screw axis: '6 sub 2'		$\frac{1}{3}$	6_2 $(3_2, 2)$
Sixfold screw axis: '6 sub 3'		$\frac{1}{2}$	6_3 $(3, 2_1)$
Sixfold screw axis: '6 sub 4'		$\frac{2}{3}$	6_4 $(3_1, 2)$
Sixfold screw axis: '6 sub 5'		$\frac{5}{6}$	6_5 $(3_2, 2_1)$
Centre of symmetry, inversion centre: '1 bar' / Reflection point, mirror point (one dimension)		None	$\bar{1}$
Inversion axis: '3 bar'		None	$\bar{3}$ $(3, \bar{1})$
Inversion axis: '4 bar'		None	$\bar{4}$ (2)
Inversion axis: '6 bar'		None	$\bar{6} \equiv 3/m$
Twofold rotation axis with centre of symmetry		None	$2/m$ $(\bar{1})$
Twofold screw axis with centre of symmetry		$\frac{1}{2}$	$2_1/m$ $(\bar{1})$
Fourfold rotation axis with centre of symmetry		None	$4/m$ $(\bar{4}, 2, \bar{1})$
'4 sub 2' screw axis with centre of symmetry		$\frac{1}{2}$	$4_2/m$ $(\bar{4}, 2, \bar{1})$
Sixfold rotation axis with centre of symmetry		None	$6/m$ $(\bar{6}, \bar{3}, 3, 2, \bar{1})$
'6 sub 3' screw axis with centre of symmetry		$\frac{1}{2}$	$6_3/m$ $(\bar{6}, \bar{3}, 3, 2_1, \bar{1})$

* Notes on the 'heights' h of symmetry points $\bar{1}, \bar{3}, \bar{4}$ and $\bar{6}$:
(1) Centres of symmetry $\bar{1}$ and $\bar{3}$, as well as inversion points $\bar{4}$ and $\bar{6}$ on $\bar{4}$ and $\bar{6}$ axes parallel to [001], occur in pairs at 'heights' h and $h + \frac{1}{2}$. In the space-group diagrams, only one fraction h is given, e.g. $\frac{1}{4}$ stands for $h = \frac{1}{4}$ and $\frac{3}{4}$. No fraction means $h = 0$ and $\frac{1}{2}$. In *cubic* space groups, however, because of their complexity, *both* fractions are given for vertical $\bar{4}$ axes, including $h = 0$ and $\frac{1}{2}$.
(2) Symmetries $4/m$ and $6/m$ contain vertical $\bar{4}$ and $\bar{6}$ axes; their $\bar{4}$ and $\bar{6}$ inversion points coincide with the centres of symmetry. This is not indicated in the space-group diagrams.
(3) Symmetries $4_2/m$ and $6_3/m$ also contain vertical $\bar{4}$ and $\bar{6}$ axes, but their $\bar{4}$ and $\bar{6}$ inversion points alternate with the centres of symmetry; i.e. $\bar{1}$ points at h and $h + \frac{1}{2}$ interleave with $\bar{4}$ or $\bar{6}$ points at $h + \frac{1}{4}$ and $h + \frac{3}{4}$. In the tetragonal and hexagonal space-group diagrams, only *one* fraction for $\bar{1}$ and one for $\bar{4}$ or $\bar{6}$ is given. In the cubic diagrams, *all four* fractions are listed for $4_2/m$; e.g. $Pm\bar{3}n$ (No. 223): $\bar{1}: 0, \frac{1}{2}; \bar{4}: \frac{1}{4}, \frac{3}{4}$.

1.4.6. Symmetry axes parallel to the plane of projection

Symmetry axis	Graphical symbol*	Screw vector of a right-handed screw rotation in units of the shortest lattice translation vector parallel to the axis	Printed symbol (partial elements in parentheses)
Twofold rotation axis		None	2
Twofold screw axis: '2 sub 1'		$\frac{1}{2}$	2_1
Fourfold rotation axis		None	4 (2)
Fourfold screw axis: '4 sub 1'		$\frac{1}{4}$	4_1 (2_1)
Fourfold screw axis: '4 sub 2'		$\frac{1}{2}$	4_2 (2)
Fourfold screw axis: '4 sub 3'		$\frac{3}{4}$	4_3 (2_1)
Inversion axis: '4 bar'		None	$\bar{4}$ (2)
Inversion point on '4 bar'-axis		–	$\bar{4}$ point

(bracket label: in cubic space groups only)

* The symbols for horizontal symmetry axes are given outside the unit cell of the space-group diagrams. *Twofold* axes always occur in pairs, at 'heights' h and $h + \frac{1}{2}$ above the plane of projection; here, a fraction h attached to such a symbol indicates two axes with heights h and $h + \frac{1}{2}$. No fraction stands for $h = 0$ and $\frac{1}{2}$. The rule of pairwise occurrence, however, is not valid for the horizontal *fourfold* axes in cubic space groups; here, *all* heights are given, including $h = 0$ and $\frac{1}{2}$. This applies also to the horizontal $\bar{4}$ axes and the $\bar{4}$ inversion points located on these axes.

1.4.7. Symmetry axes inclined to the plane of projection (in cubic space groups only)

Symmetry axis	Graphical symbol*		Screw vector of a right-handed screw rotation in units of the shortest lattice translation vector parallel to the axis	Printed symbol (partial elements in parentheses)
Twofold rotation axis		Parallel to a face diagonal of the cube	None	2
Twofold screw axis: '2 sub 1'			$\frac{1}{2}$	2_1
Threefold rotation axis		Parallel to a body diagonal of the cube	None	3
Threefold screw axis: '3 sub 1'			$\frac{1}{3}$	3_1
Threefold screw axis: '3 sub 2'			$\frac{2}{3}$	3_2
Inversion axis: '3 bar'			None	$\bar{3}$ $(3, \bar{1})$

* The dots mark the intersection points of axes with the plane at $h = 0$. In some cases, the intersection points are obscured by symbols of symmetry elements with height $h \geq 0$; examples: $Fd\bar{3}$ (203), origin choice 2; $Pn\bar{3}n$ (222), origin choice 2; $Pm\bar{3}n$ (223); $Im\bar{3}m$ (229); $Ia\bar{3}d$ (230).

References

1.2

Internationale Tabellen zur Bestimmung von Kristallstrukturen (1935). I. Band, edited by C. Hermann. Berlin: Borntraeger. [Reprint with corrections: Ann Arbor: Edwards (1944). Abbreviated as *IT* (1935).]

International Tables for X-ray Crystallography (1952). Vol. I, edited by N. F. M. Henry & K. Lonsdale. Birmingham: Kynoch Press. [Abbreviated as *IT* (1952).]

Wolff, P. M. de, Belov, N. V., Bertaut, E. F., Buerger, M. J., Donnay, J. D. H., Fischer, W., Hahn, Th., Koptsik, V. A., Mackay, A. L., Wondratschek, H., Wilson, A. J. C. & Abrahams, S. C. (1985). *Nomenclature for crystal families, Bravais-lattice types and arithmetic classes. Report of the International Union of Crystallography Ad-hoc Committee on the Nomenclature of Symmetry. Acta Cryst.* A**41**, 278–280.

1.3

Flack, H. D., Wondratschek, H., Hahn, Th. & Abrahams, S. C. (2000). *Symmetry elements in space groups and point groups. Addenda to two IUCr Reports on the Nomenclature of Symmetry. Acta Cryst.* A**56**, 96–98.

Wolff, P. M. de, Billiet, Y., Donnay, J. D. H., Fischer, W., Galiulin, R. B., Glazer, A. M., Senechal, M., Shoemaker, D. P., Wondratschek, H., Hahn, Th., Wilson, A. J. C. & Abrahams, S. C. (1989). *Definition of symmetry elements in space groups and point groups. Report of the International Union of Crystallography Ad-hoc Committee on the Nomenclature of Symmetry. Acta Cryst.* A**45**, 494–499.

Wolff, P. M. de, Billiet, Y., Donnay, J. D. H., Fischer, W., Galiulin, R. B., Glazer, A. M., Hahn, Th., Senechal, M., Shoemaker, D. P., Wondratschek, H., Wilson, A. J. C. & Abrahams, S. C. (1992). *Symbols for symmetry elements and symmetry operations. Final Report of the International Union of Crystallography Ad-hoc Committee on the Nomenclature of Symmetry. Acta Cryst.* A**48**, 727–732.

1.4

Internationale Tabellen zur Bestimmung von Kristallstrukturen (1935). I. Band, edited by C. Hermann. Berlin: Borntraeger. [Reprint with corrections: Ann Arbor: Edwards (1944). Abbreviated as *IT* (1935).]

International Tables for X-ray Crystallography (1952). Vol. I, edited by N. F. M. Henry & K. Lonsdale. Birmingham: Kynoch Press. [Abbreviated as *IT* (1952).]

Wolff, P. M. de, Billiet, Y., Donnay, J. D. H., Fischer, W., Galiulin, R. B., Glazer, A. M., Hahn, Th., Senechal, M., Shoemaker, D. P., Wondratschek, H., Wilson, A. J. C. & Abrahams, S. C. (1992). *Symbols for symmetry elements and symmetry operations. Final Report of the International Union of Crystallography Ad-hoc Committee on the Nomenclature of Symmetry. Acta Cryst.* A**48**, 727–732.

2. GUIDE TO THE USE OF THE SPACE-GROUP TABLES

By Th. Hahn and A. Looijenga-Vos

Based on contributions by H. Arnold, M. J. Buerger, P. M. de Wolff, J. D. H. Donnay, W. Fischer, Th. Hahn, G. A. Langlet, A. Looijenga-Vos and H. Wondratschek

2.1. Classification and coordinate systems of space groups

BY TH. HAHN AND A. LOOIJENGA-VOS

2.1.1. Introduction

The present volume is a computer-based extension and complete revision of the symmetry tables of the two previous series of *International Tables*, the *Internationale Tabellen zur Bestimmung von Kristallstrukturen* (1935) and the *International Tables for X-ray Crystallography* (1952).*

The main part of the volume consists of tables and diagrams for the 17 types of plane groups (Part 6) and the 230 types of space groups (Part 7). The two types of line groups are treated separately in Section 2.2.17, because of their simplicity. For the history of the *Tables* and a comparison of the various editions, reference is made to the *Preface* of this volume. Attention is drawn to Part 1 where the symbols and terms used in this volume are defined.

The present part forms a *guide* to the entries in the space-group tables with instructions for their practical use. Only a minimum of theory is provided, and the emphasis is on practical aspects. For the theoretical background the reader is referred to Parts 8–15, which include also suitable references. A textbook version of space-group symmetry and the use of these tables (with exercises) is provided by Hahn & Wondratschek (1994).

2.1.2. Space-group classification

In this volume, the plane groups and space groups are classified according to three criteria:

(i) According to *geometric crystal classes, i.e.* according to the crystallographic point group to which a particular space group belongs. There are 10 crystal classes in two dimensions and 32 in three dimensions. They are described and listed in Part 10 and in column 4 of Table 2.1.2.1. [For arithmetic crystal classes, see Section 8.2.3 in this volume and Chapter 1.4 of *International Tables for Crystallography*, Vol. C (2004).]

(ii) According to *crystal families*. The term crystal family designates the classification of the 17 plane groups into four categories and of the 230 space groups into *six* categories, as displayed in column 1 of Table 2.1.2.1. Here all 'hexagonal', 'trigonal' and 'rhombohedral' space groups are contained in one family, the hexagonal crystal family. The 'crystal family' thus corresponds to the term 'crystal system', as used frequently in the American and Russian literature.

The crystal families are symbolized by the lower-case letters *a, m, o, t, h, c,* as listed in column 2 of Table 2.1.2.1. If these letters are combined with the appropriate capital letters for the lattice-centring types (*cf.* Chapter 1.2), symbols for the 14 Bravais lattices result. These symbols and their occurrence in the crystal families are shown in column 8 of Table 2.1.2.1; *mS* and *oS* are the standard setting-independent symbols for the centred monoclinic and the one-face centred orthorhombic Bravais lattices, *cf.* de Wolff *et al.* (1985); symbols between parentheses represent alternative settings of these Bravais lattices.

(iii) According to *crystal systems*. This classification collects the plane groups into four categories and the space groups into *seven* categories. The classifications according to crystal families and crystal systems are the same for two dimensions.

For three dimensions, this applies to the triclinic, monoclinic, orthorhombic, tetragonal and cubic systems. The only complication exists in the hexagonal crystal family for which several subdivisions

into systems have been proposed in the literature. In this volume, as well as in *IT* (1952), the space groups of the hexagonal crystal family are grouped into two 'crystal systems' as follows: all space groups belonging to the five crystal classes 3, $\bar{3}$, 32, 3*m* and $\bar{3}m$, *i.e.* having 3, 3_1, 3_2 or $\bar{3}$ as principal axis, form the *trigonal* crystal system, irrespective of whether the Bravais lattice is *hP* or *hR*; all space groups belonging to the seven crystal classes 6, $\bar{6}$, 6/*m*, 622, 6*mm*, $\bar{6}2m$ and 6/*mmm*, *i.e.* having 6, 6_1, 6_2, 6_3, 6_4, 6_5 or $\bar{6}$ as principal axis, form the *hexagonal* crystal system; here the lattice is always *hP* (*cf.* Section 8.2.8). The crystal systems, as defined above, are listed in column 3 of Table 2.1.2.1.

A different subdivision of the hexagonal crystal family is in use, mainly in the French literature. It consists of grouping all space groups based on the hexagonal Bravais lattice *hP* (lattice point symmetry 6/*mmm*) into the 'hexagonal' system and all space groups based on the rhombohedral Bravais lattice *hR* (lattice point symmetry $\bar{3}m$) into the 'rhombohedral' system. In Section 8.2.8, these systems are called 'Lattice systems'. They were called 'Bravais systems' in earlier editions of this volume.

The theoretical background for the classification of space groups is provided in Chapter 8.2.

2.1.3. Conventional coordinate systems and cells

A plane group or space group usually is described by means of a *crystallographic coordinate system*, consisting of a *crystallographic basis* (basis vectors are lattice vectors) and a *crystallographic origin* (origin at a centre of symmetry or at a point of high site symmetry). The choice of such a coordinate system is not mandatory since in principle a crystal structure can be referred to any coordinate system; *cf.* Section 8.1.4.

The selection of a crystallographic coordinate system is not unique. Conventionally, a right-handed set of basis vectors is taken such that the symmetry of the plane or space group is displayed best. With this convention, which is followed in the present volume, the specific restrictions imposed on the cell parameters by each crystal family become particularly simple. They are listed in columns 6 and 7 of Table 2.1.2.1. If within these restrictions the smallest cell is chosen, a *conventional* (crystallographic) *basis* results. Together with the selection of an appropriate *conventional* (crystallographic) *origin* (*cf.* Sections 2.2.2 and 2.2.7), such a basis defines a *conventional* (crystallographic) *coordinate system* and a *conventional cell*. The conventional cell of a point lattice or a space group, obtained in this way, turns out to be either *primitive* or to exhibit one of the *centring types* listed in Chapter 1.2. The centring type of a conventional cell is transferred to the lattice which is described by this cell; hence, we speak of primitive, face-centred, body-centred *etc.* lattices. Similarly, the cell parameters are often called lattice parameters; *cf.* Section 8.3.1 and Chapter 9.1 for further details.

In the triclinic, monoclinic and orthorhombic crystal systems, additional conventions (for instance cell reduction or metrical conventions based on the lengths of the cell edges) are needed to determine the choice and the labelling of the axes. Reduced bases are treated in Chapters 9.1 and 9.2, orthorhombic settings in Section 2.2.6.4, and monoclinic settings and cell choices in Section 2.2.16.

In this volume, all space groups within a crystal family are referred to the same kind of conventional coordinate system with the exception of the hexagonal crystal family in three dimensions. Here, two kinds of coordinate systems are used, the hexagonal and the rhombohedral systems. In accordance with common crystallographic practice, all space groups based on the hexagonal Bravais lattice *hP* (18 trigonal and 27 hexagonal space groups) are described

---

* Throughout this volume, these editions are abbreviated as *IT* (1935) and *IT* (1952).

Table 2.1.2.1. *Crystal families, crystal systems, conventional coordinate systems and Bravais lattices in one, two and three dimensions*

Crystal family	Symbol*	Crystal system	Crystallographic point groups†	No. of space groups	Conventional coordinate system		Bravais lattices*
					Restrictions on cell parameters	Parameters to be determined	
One dimension							
–	–	–	$1, \boxed{m}$	2	None	a	p
Two dimensions							
Oblique (monoclinic)	m	Oblique	$1, \boxed{2}$	2	None	a, b γ‡	mp
Rectangular (orthorhombic)	o	Rectangular	$m, \boxed{2mm}$	7	$\gamma = 90°$	a, b	op oc
Square (tetragonal)	t	Square	$\boxed{4}, \boxed{4mm}$	3	$a = b$ $\gamma = 90°$	a	tp
Hexagonal	h	Hexagonal	$3, \boxed{6}$ $3m, \boxed{6mm}$	5	$a = b$ $\gamma = 120°$	a	hp
Three dimensions							
Triclinic (anorthic)	a	Triclinic	$1, \boxed{\bar{1}}$	2	None	$a, b, c,$ α, β, γ	aP
Monoclinic	m	Monoclinic	$2, m, \boxed{2/m}$	13	b-unique setting $\alpha = \gamma = 90°$	a, b, c β‡	mP $mS\,(mC, mA, mI)$
					c-unique setting $\alpha = \beta = 90°$	$a, b, c,$ γ‡	mP $mS\,(mA, mB, mI)$
Orthorhombic	o	Orthorhombic	$222, mm2, \boxed{mmm}$	59	$\alpha = \beta = \gamma = 90°$	a, b, c	oP $oS\,(oC, oA, oB)$ oI oF
Tetragonal	t	Tetragonal	$4, \bar{4}, \boxed{4/m}$ $422, 4mm, \bar{4}2m,$ $\boxed{4/mmm}$	68	$a = b$ $\alpha = \beta = \gamma = 90°$	a, c	tP tI
Hexagonal	h	Trigonal	$3, \boxed{\bar{3}}$ $32, 3m, \boxed{\bar{3}m}$	18	$a = b$ $\alpha = \beta = 90°, \ \gamma = 120°$	a, c	hP
				7	$a = b = c$ $\alpha = \beta = \gamma$ (rhombohedral axes, primitive cell) $a = b$ $\alpha = \beta = 90°, \gamma = 120°$ (hexagonal axes, triple obverse cell)	a, α	hR
		Hexagonal	$6, \bar{6}, \boxed{6/m}$ $622, 6mm, \bar{6}2m,$ $\boxed{6/mmm}$	27	$a = b$ $\alpha = \beta = 90°, \gamma = 120°$	a, c	hP
Cubic	c	Cubic	$23, \boxed{m\bar{3}}$ $432, \bar{4}3m, \boxed{m\bar{3}m}$	36	$a = b = c$ $\alpha = \beta = \gamma = 90°$	a	cP cI cF

* The symbols for crystal families (column 2) and Bravais lattices (column 8) were adopted by the International Union of Crystallography in 1985; *cf.* de Wolff *et al.* (1985).
† Symbols surrounded by dashed or full lines indicate Laue groups; full lines indicate Laue groups which are also lattice point symmetries (holohedries).
‡ These angles are conventionally taken to be non-acute, *i.e.* $\geq 90°$.

only with a hexagonal coordinate system (primitive cell),* whereas the seven space groups based on the rhombohedral Bravais lattice hR are treated in two versions, one referred to 'hexagonal axes' (triple obverse cell) and one to 'rhombohedral axes' (primitive cell); *cf.* Chapter 1.2. In practice, hexagonal axes are preferred because they are easier to visualize.

Note: For convenience, the relations between the cell parameters a, c of the triple hexagonal cell and the cell parameters a', α' of the primitive rhombohedral cell (*cf.* Table 2.1.2.1) are listed:

$$a = a'\sqrt{2}\sqrt{1 - \cos\alpha'} = 2a'\sin\frac{\alpha'}{2}$$

$$c = a'\sqrt{3}\sqrt{1 + 2\cos\alpha'}$$

$$\frac{c}{a} = \sqrt{\frac{3}{2}}\sqrt{\frac{1 + 2\cos\alpha'}{1 - \cos\alpha'}} = \sqrt{\frac{9}{4\sin^2(\alpha'/2)} - 3}$$

$$a' = \tfrac{1}{3}\sqrt{3a^2 + c^2}$$

$$\sin\frac{\alpha'}{2} = \frac{3}{2\sqrt{3 + (c^2/a^2)}} \quad \text{or} \quad \cos\alpha' = \frac{(c^2/a^2) - \tfrac{3}{2}}{(c^2/a^2) + 3}.$$

---

* For a rhombohedral description (D cell) of the hexagonal Bravais lattice see Section 4.3.5.3.

2.2. Contents and arrangement of the tables

By Th. Hahn and A. Looijenga-Vos

2.2.1. General layout

The presentation of the plane-group and space-group data in Parts 6 and 7 follows the style of the previous editions of *International Tables*. The entries for a space group are printed on two facing pages as shown below; an example (*Cmm*2, No. 35) is provided inside the front and back covers. Deviations from this standard sequence (mainly for cubic space groups) are indicated on the relevant pages.

Left-hand page:

(1)   *Headline*

(2)   *Diagrams* for the symmetry elements and the general position (for graphical symbols of symmetry elements see Chapter 1.4)

(3)   *Origin*

(4)   *Asymmetric unit*

(5)   *Symmetry operations*

Right-hand page:

(6)   *Headline* in abbreviated form

(7)   *Generators selected*; this information is the basis for the order of the entries under *Symmetry operations* and *Positions*

(8)   General and special *Positions*, with the following columns:

   Multiplicity

   Wyckoff letter

   Site symmetry, given by the oriented site-symmetry symbol

   Coordinates

   Reflection conditions

   Note: In a few space groups, two special positions with the same reflection conditions are printed on the same line

(9)   *Symmetry of special projections* (not given for plane groups)

(10)   *Maximal non-isomorphic subgroups*

(11)   *Maximal isomorphic subgroups of lowest index*

(12)   *Minimal non-isomorphic supergroups*

Note: Symbols for *Lattice complexes* of the plane groups and space groups are given in Tables 14.2.3.1 and 14.2.3.2. Normalizers of space groups are listed in Part 15.

2.2.2. Space groups with more than one description

For several space groups, more than one description is available. Three cases occur:

(i) *Two choices of origin (cf. Section 2.2.7)*

For all centrosymmetric space groups, the tables contain a description with a centre of symmetry as origin. Some centrosymmetric space groups, however, contain points of high site symmetry that do not coincide with a centre of symmetry. For these 24 cases, a further description (including diagrams) with a high-symmetry point as origin is provided. Neither of the two origin choices is considered standard. Noncentrosymmetric space groups and all plane groups are described with only one choice of origin.

Examples

(1) *Pnnn* (48)

   Origin choice 1 at a point with site symmetry 222

   Origin choice 2 at a centre with site symmetry $\bar{1}$.

(2) *Fd$\bar{3}$m* (227)

   Origin choice 1 at a point with site symmetry $\bar{4}3m$

   Origin choice 2 at a centre with site symmetry $\bar{3}m$.

(ii) *Monoclinic space groups*

Two complete descriptions are given for each of the 13 monoclinic space groups, one for the setting with 'unique axis *b*', followed by one for the setting with 'unique axis *c*'.

Additional descriptions in synoptic form are provided for the following eight monoclinic space groups with centred lattices or glide planes:

   *C*2 (5), *Pc* (7), *Cm* (8), *Cc* (9), *C*2/*m* (12), *P*2/*c* (13), *P*2$_1$/*c* (14), *C*2/*c* (15).

These synoptic descriptions consist of abbreviated treatments for three 'cell choices', here called 'cell choices 1, 2 and 3'. Cell choice 1 corresponds to the complete treatment, mentioned above; for comparative purposes, it is repeated among the synoptic descriptions which, for each setting, are printed on two facing pages. The cell choices and their relations are explained in Section 2.2.16.

(iii) *Rhombohedral space groups*

The seven rhombohedral space groups *R*3 (146), *R*$\bar{3}$ (148), *R*32 (155), *R*3*m* (160), *R*3*c* (161), *R*$\bar{3}$*m* (166), and *R*$\bar{3}$*c* (167) are described with two coordinate systems, first with *hexagonal axes* (triple hexagonal cell) and second with *rhombohedral axes* (primitive rhombohedral cell). For both descriptions, the same space-group symbol is used. The relations between the cell parameters of the two cells are listed in Chapter 2.1.

The hexagonal triple cell is given in the *obverse* setting (centring points $\frac{2}{3}, \frac{1}{3}, \frac{1}{3}; \frac{1}{3}, \frac{2}{3}, \frac{2}{3}$). In *IT* (1935), the *reverse* setting (centring points $\frac{1}{3}, \frac{2}{3}, \frac{1}{3}; \frac{2}{3}, \frac{1}{3}, \frac{2}{3}$) was employed; *cf.* Chapter 1.2.

2.2.3. Headline

The description of each plane group or space group starts with a headline on a left-hand page, consisting of two (sometimes three) lines which contain the following information, when read from left to right.

First line

(1) The *short international* (Hermann–Mauguin) *symbol* for the plane or space group. These symbols will be further referred to as Hermann–Mauguin symbols. A detailed discussion of space-group symbols is given in Chapter 12.2, a brief summary in Section 2.2.4.

   Note on standard monoclinic space-group symbols: In order to facilitate recognition of a monoclinic space-group type, the familiar short symbol for the *b*-axis setting (*e.g. P*2$_1$/*c* for No. 14 or *C*2/*c* for No. 15) has been adopted as the *standard symbol* for a space-group type. It appears in the headline of *every description of this space group* and thus does not carry any information about the setting or the cell choice of this particular description. No other short symbols for monoclinic space groups are used in this volume (*cf.* Section 2.2.16).

(2) The *Schoenflies symbol* for the space group.

   Note: No Schoenflies symbols exist for the plane groups.

(3) The *short international* (Hermann–Mauguin) *symbol* for the point group to which the plane or space group belongs (*cf.* Chapter 12.1).

(4) The name of the *crystal system* (*cf.* Table 2.1.2.1).

Second line

(5) The sequential *number of the plane or space group*, as introduced in *IT* (1952).

(6) The *full international* (Hermann–Mauguin) *symbol* for the plane or space group.

For monoclinic space groups, the headline of every description contains the full symbol appropriate to that description.

(7) The *Patterson symmetry* (see Section 2.2.5).

Third line

This line is used, where appropriate, to indicate origin choices, settings, cell choices and coordinate axes (see Section 2.2.2). For five orthorhombic space groups, an entry 'Former space-group symbol' is given; *cf.* Chapter 1.3, Note (x).

2.2.4. International (Hermann–Mauguin) symbols for plane groups and space groups (*cf.* Chapter 12.2)

2.2.4.1. *Present symbols*

Both the short and the full Hermann–Mauguin symbols consist of two parts: (i) a letter indicating the centring type of the conventional cell, and (ii) a set of characters indicating symmetry elements of the space group (modified point-group symbol).

(i) The letters for the centring types of cells are listed in Chapter 1.2. Lower-case letters are used for two dimensions (nets), capital letters for three dimensions (lattices).

(ii) The one, two or three entries after the centring letter refer to the one, two or three kinds of *symmetry directions* of the lattice belonging to the space group. These symmetry directions were called *blickrichtungen* by Heesch (1929). Symmetry directions occur either as singular directions (as in the monoclinic and orthorhombic crystal systems) or as sets of symmetrically equivalent symmetry directions (as in the higher-symmetrical crystal systems). Only one representative of each set is required. The (sets of) symmetry directions and their sequence for the different lattices are summarized in Table 2.2.4.1. According to their position in this sequence, the symmetry directions are referred to as 'primary', 'secondary' and 'tertiary' directions.

This sequence of lattice symmetry directions is transferred to the sequence of positions in the corresponding Hermann–Mauguin space-group symbols. Each position contains one or two characters designating symmetry elements (axes and planes) of the space group (*cf.* Chapter 1.3) that occur for the corresponding lattice symmetry direction. Symmetry planes are represented by their normals; if a symmetry axis and a normal to a symmetry plane are parallel, the two characters (symmetry symbols) are separated by a slash, as in $P6_3/m$ or $P2/m$ ('two over *m*').

For the different crystal lattices, the Hermann–Mauguin space-group symbols have the following form:

(i) *Triclinic* lattices have no symmetry direction because they have, in addition to translations, only centres of symmetry, $\bar{1}$. Thus, only two triclinic space groups, $P1$ (1) and $P\bar{1}$ (2), exist.

(ii) *Monoclinic* lattices have one symmetry direction. Thus, for monoclinic space groups, only one position after the centring letter is needed. This is used in the *short* Hermann–Mauguin symbols, as in $P2_1$. Conventionally, the symmetry direction is labelled either *b* ('unique axis *b*') or *c* ('unique axis *c*').

In order to distinguish between the different settings, the *full* Hermann–Mauguin symbol contains two extra entries '1'. They indicate those two axial directions that are not symmetry directions

Table 2.2.4.1. *Lattice symmetry directions for two and three dimensions*

Directions that belong to the same set of equivalent symmetry directions are collected between braces. The first entry in each set is taken as the representative of that set.

| Lattice | Symmetry direction (position in Hermann–Mauguin symbol) | | |
	Primary	Secondary	Tertiary
Two dimensions			
Oblique	Rotation point in plane		
Rectangular		[10]	[01]
Square		$\left\{\begin{array}{c}[10]\\ [01]\end{array}\right\}$	$\left\{\begin{array}{c}[1\bar{1}]\\ [11]\end{array}\right\}$
Hexagonal		$\left\{\begin{array}{c}[10]\\ [01]\\ [\bar{1}\bar{1}]\end{array}\right\}$	$\left\{\begin{array}{c}[1\bar{1}]\\ [12]\\ [\bar{2}\bar{1}]\end{array}\right\}$
Three dimensions			
Triclinic	None		
Monoclinic*	[010] ('unique axis *b*') [001] ('unique axis *c*')		
Orthorhombic	[100]	[010]	[001]
Tetragonal	[001]	$\left\{\begin{array}{c}[100]\\ [010]\end{array}\right\}$	$\left\{\begin{array}{c}[1\bar{1}0]\\ [110]\end{array}\right\}$
Hexagonal	[001]	$\left\{\begin{array}{c}[100]\\ [010]\\ [\bar{1}\bar{1}0]\end{array}\right\}$	$\left\{\begin{array}{c}[1\bar{1}0]\\ [120]\\ [\bar{2}\bar{1}0]\end{array}\right\}$
Rhombohedral (hexagonal axes)	[001]	$\left\{\begin{array}{c}[100]\\ [010]\\ [\bar{1}\bar{1}0]\end{array}\right\}$	
Rhombohedral (rhombohedral axes)	[111]	$\left\{\begin{array}{c}[1\bar{1}0]\\ [01\bar{1}]\\ [\bar{1}01]\end{array}\right\}$	
Cubic	$\left\{\begin{array}{c}[100]\\ [010]\\ [001]\end{array}\right\}$	$\left\{\begin{array}{c}[111]\\ [1\bar{1}1]\\ [\bar{1}1\bar{1}]\\ [\bar{1}\bar{1}1]\end{array}\right\}$	$\left\{\begin{array}{cc}[1\bar{1}0] & [110]\\ [01\bar{1}] & [011]\\ [\bar{1}01] & [101]\end{array}\right\}$

* For the full Hermann–Mauguin symbols see Section 2.2.4.1.

of the lattice. Thus, the symbols $P121$, $P112$ and $P211$ show that the *b* axis, *c* axis and *a* axis, respectively, is the unique axis. Similar considerations apply to the three *rectangular* plane groups *pm*, *pg* and *cm* (*e.g.* plane group No. 5: short symbol *cm*, full symbol *c*1*m*1 or *c*11*m*).

(iii) *Rhombohedral* lattices have two kinds of symmetry directions. Thus, the symbols of the seven rhombohedral space groups contain only two entries after the letter *R*, as in $R3m$ or $R3c$.

(iv) *Orthorhombic, tetragonal, hexagonal* and *cubic* lattices have three kinds of symmetry directions. Hence, the corresponding space-group symbols have three entries after the centring letter, as in $Pmna$, $P3m1$, $P6cc$ or $Ia\bar{3}d$.

Lattice symmetry directions that carry no symmetry elements for the space group under consideration are represented by the symbol '1', as in $P3m1$ and $P31m$. If no misinterpretation is possible, entries '1' at the end of a space-group symbol are omitted, as in $P6$ (instead of $P611$), $R\bar{3}$ (instead of $R\bar{3}1$), $I4_1$ (instead of $I4_111$), $F23$ (instead of $F231$); similarly for the plane groups.

Short and *full* Hermann–Mauguin symbols differ only for the plane groups of class *m*, for the monoclinic space groups, and for the space groups of crystal classes *mmm*, 4/*mmm*, $\bar{3}m$, 6/*mmm*, $m\bar{3}$ and $m\bar{3}m$. In the full symbols, symmetry axes *and* symmetry planes for each symmetry direction are listed; in the short symbols, symmetry axes are suppressed as much as possible. Thus, for space group No. 62, the full symbol is $P2_1/n\,2_1/m\,2_1/a$ and the short symbol is *Pnma*. For No. 194, the full symbol is $P6_3/m\,2/m\,2/c$ and the short symbol is $P6_3/mmc$. For No. 230, the two symbols are $I4_1/a\,\bar{3}\,2/d$ and $Ia\bar{3}d$.

Many space groups contain more kinds of symmetry elements than are indicated in the full symbol ('additional symmetry elements', *cf.* Chapter 4.1). A complete listing of the symmetry elements is given in Tables 4.2.1.1 and 4.3.2.1 under the heading *Extended full symbols*. Note that a centre of symmetry is never explicitly indicated (except for space group $P\bar{1}$); its presence or absence, however, can be readily inferred from the space-group symbol.

2.2.4.2. *Changes in Hermann–Mauguin space-group symbols as compared with the 1952 and 1935 editions of International Tables*

Extensive changes in the space-group symbols were applied in *IT* (1952) as compared with the original Hermann–Mauguin symbols of *IT* (1935), especially in the tetragonal, trigonal and hexagonal crystal systems. Moreover, new symbols for the *c*-axis setting of monoclinic space groups were introduced. All these changes are recorded on pp. 51 and 543–544 of *IT* (1952). In the present edition, the symbols of the 1952 edition are retained, except for the following four cases (*cf.* Chapter 12.4).

(i) *Two-dimensional groups*
Short Hermann–Mauguin symbols differing from the corresponding full symbols in *IT* (1952) are replaced by the full symbols for the listed plane groups in Table 2.2.4.2.

For the two-dimensional point group with two mutually perpendicular mirror lines, the symbol *mm* is changed to 2*mm*.

For plane group No. 2, the entries '1' at the end of the full symbol are omitted:

No. 2: Change from *p*211 to *p*2.

With these changes, the symbols of the two-dimensional groups follow the rules that were introduced in *IT* (1952) for the space groups.

(ii) *Monoclinic space groups*
Additional *full* Hermann–Mauguin symbols are introduced for the eight monoclinic space groups with centred lattices or glide planes (Nos. 5, 7–9, 12–15) to indicate the various settings and cell choices. A complete list of symbols, including also the *a*-axis

setting, is contained in Table 4.3.2.1; further details are given in Section 2.2.16.

For standard *short* monoclinic space-group symbols see Sections 2.2.3 and 2.2.16.

(iii) *Cubic groups*
The short symbols for all space groups belonging to the two cubic crystal classes $m\bar{3}$ and $m\bar{3}m$ now contain the symbol $\bar{3}$ instead of 3. This applies to space groups Nos. 200–206 and 221–230, as well as to the two point groups $m\bar{3}$ and $m\bar{3}m$.

Examples
No. 205: Change from *Pa*3 to *Pa*$\bar{3}$
No. 230: Change from *Ia*3*d* to *Ia*$\bar{3}$*d*.

With this change, the centrosymmetric nature of these groups is apparent also in the short symbols.

(iv) *Glide-plane symbol e*
For the recent introduction of the 'double glide plane' *e* into five space-group symbols, see Chapter 1.3, Note (x).

2.2.5. Patterson symmetry

The entry *Patterson symmetry* in the headline gives the space group of the *Patterson function* $P(x, y, z)$. With neglect of anomalous dispersion, this function is defined by the formula

$$P(x, y, z) = \frac{1}{V} \sum_h \sum_k \sum_l |F(hkl)|^2 \cos 2\pi(hx + ky + lz).$$

The Patterson function represents the convolution of a structure with its inverse or the pair-correlation function of a structure. A detailed discussion of its use for structure determination is given by Buerger (1959). The space group of the Patterson function is identical to that of the 'vector set' of the structure, and is thus always centrosymmetric and symmorphic.*

The symbol for the Patterson space group of a crystal structure can be deduced from that of its space group in two steps:

(i) Glide planes and screw axes have to be replaced by the corresponding mirror planes and rotation axes, resulting in a symmorphic space group.

(ii) If this symmorphic space group is not centrosymmetric, inversions have to be added.

There are 7 different Patterson symmetries in two dimensions and 24 in three dimensions. They are listed in Table 2.2.5.1. Account is taken of the fact that the Laue class $\bar{3}m$ combines in two ways with the hexagonal translation lattice, namely as $\bar{3}m1$ and as $\bar{3}1m$.

Note: For the four orthorhombic space groups with *A* cells (Nos. 38–41), the standard symbol for their Patterson symmetry, *Cmmm*, is added (between parentheses) after the actual symbol *Ammm* in the space-group tables.

The 'point group part' of the symbol of the Patterson symmetry represents the *Laue class* to which the plane group or space group belongs (*cf.* Table 2.1.2.1). In the absence of anomalous dispersion, the Laue class of a crystal expresses the *point symmetry of its diffraction record*, *i.e.* the symmetry of the reciprocal lattice weighted with *I*(*hkl*).

Table 2.2.4.2. *Changes in Hermann–Mauguin symbols for two-dimensional groups*

No.	*IT* (1952)	Present edition
6	*pmm*	*p*2*mm*
7	*pmg*	*p*2*mg*
8	*pgg*	*p*2*gg*
9	*cmm*	*c*2*mm*
11	*p*4*m*	*p*4*mm*
12	*p*4*g*	*p*4*gm*
17	*p*6*m*	*p*6*mm*

---

* A space group is called 'symmorphic' if, apart from the lattice translations, all *generating* symmetry operations leave one common point fixed. Permitted as generators are thus only the point-group operations: rotations, reflections, inversions and rotoinversions (*cf.* Section 8.1.6).

Table 2.2.5.1. *Patterson symmetries for two and three dimensions*

Laue class	Lattice type				Patterson symmetry (with space-group number)			
Two dimensions								
2	p				$p2$ (2)			
$2mm$	p	c			$p2mm$ (6)	$c2mm$ (9)		
4	p				$p4$ (10)			
$4mm$	p				$p4mm$ (11)			
6	p				$p6$ (16)			
$6mm$	p				$p6mm$ (17)			
Three dimensions								
$\bar{1}$	P				$P\bar{1}$ (2)			
$2/m$	P	C			$P2/m$ (10)	$C2/m$ (12)		
mmm	P	C	I	F	$Pmmm$ (47)	$Cmmm$ (65)	$Immm$ (71)	$Fmmm$ (69)
$4/m$	P		I		$P4/m$ (83)		$I4/m$ (87)	
$4/mmm$	P		I		$P4/mmm$ (123)		$I4/mmm$ (139)	
$\bar{3}$	P			R	$P\bar{3}$ (147)			$R\bar{3}$ (148)
$\left\{\begin{array}{c}\bar{3}m1\\ \bar{3}1m\end{array}\right.$ $\bar{3}m1$	P			R	$P\bar{3}m1$ (164)			$R\bar{3}m$ (166)
$\bar{3}1m$	P				$P\bar{3}1m$ (162)			
$6/m$	P				$P6/m$ (175)			
$6/mmm$	P				$P6/mmm$ (191)			
$m\bar{3}$	P	I	F		$Pm\bar{3}$ (200)		$Im\bar{3}$ (204)	$Fm\bar{3}$ (202)
$m\bar{3}m$	P	I	F		$Pm\bar{3}m$ (221)		$Im\bar{3}m$ (229)	$Fm\bar{3}m$ (225)

2.2.6. Space-group diagrams

The space-group diagrams serve two purposes: (i) to show the relative locations and orientations of the symmetry elements and (ii) to illustrate the arrangement of a set of symmetrically equivalent points of the general position.

All diagrams are orthogonal projections, *i.e.* the projection direction is perpendicular to the plane of the figure. Apart from the descriptions of the rhombohedral space groups with 'rhombohedral axes' (*cf.* Section 2.2.6.6), the projection direction is always a cell axis. If other axes are not parallel to the plane of the figure, they are indicated by the subscript p, as a_p, b_p or c_p. This applies to one or two axes for triclinic and monoclinic space groups (*cf.* Figs. 2.2.6.1 to 2.2.6.3), as well as to the three rhombohedral axes in Fig. 2.2.6.9.

The graphical symbols for symmetry elements, as used in the drawings, are displayed in Chapter 1.4.

In the diagrams, 'heights' h above the projection plane are indicated for symmetry planes and symmetry axes *parallel* to the projection plane, as well as for centres of symmetry. The heights are given as fractions of the shortest lattice translation normal to the projection plane and, if different from 0, are printed next to the graphical symbols. Each symmetry element at height h is accompanied by another symmetry element of the same type at height $h + \frac{1}{2}$ (this does not apply to the horizontal fourfold axes in the cubic diagrams). In the space-group diagrams, only the symmetry element at height h is indicated (*cf.* Chapter 1.4).

Schematic representations of the diagrams, displaying the origin, the labels of the axes, and the projection direction $[uvw]$, are given in Figs. 2.2.6.1 to 2.2.6.10 (except Fig. 2.2.6.6). The general-position diagrams are indicated by the letter G.

2.2.6.1. *Plane groups*

Each description of a plane group contains two diagrams, one for the symmetry elements (left) and one for the general position (right). The two axes are labelled a and b, with a pointing downwards and b running from left to right.

2.2.6.2. *Triclinic space groups*

For each of the two triclinic space groups, three elevations (along a, b and c) are given, in addition to the general-position diagram G (projected along c) at the lower right of the set, as illustrated in Fig. 2.2.6.1.

The diagrams represent a reduced cell of type II for which the three interaxial angles are non-acute, *i.e.* $\alpha, \beta, \gamma \geq 90°$. For a cell of type I, all angles are acute, *i.e.* $\alpha, \beta, \gamma < 90°$. For a discussion of the two types of reduced cells, reference is made to Section 9.2.2.

2.2.6.3. *Monoclinic space groups (cf. Sections 2.2.2 and 2.2.16)*

The 'complete treatment' of each of the two settings contains four diagrams (Figs. 2.2.6.2 and 2.2.6.3). Three of them are projections of the symmetry elements, taken along the unique axis (upper left) and along the other two axes (lower left and upper right). For the general position, only the projection along the unique axis is given (lower right).

The 'synoptic descriptions' of the three cell choices (for each setting) are headed by a pair of diagrams, as illustrated in Fig. 2.2.6.4. The drawings on the left display the symmetry elements and the ones on the right the general position (labelled G). Each diagram is a projection of four neighbouring unit cells along the unique axis. It contains the outlines of the three cell choices drawn as heavy lines. For the labelling of the axes, see Fig. 2.2.6.4. The headline of the description of each cell choice contains a small-scale drawing, indicating the basis vectors and the cell that apply to that description.

2.2.6.4. *Orthorhombic space groups and orthorhombic settings*

The space-group tables contain a set of four diagrams for each orthorhombic space group. The set consists of three projections of the symmetry elements [along the c axis (upper left), the a axis (lower left) and the b axis (upper right)] in addition to the general-position diagram, which is given only in the projection along c

Table 2.2.6.1. *Numbers of distinct projections and different Hermann–Mauguin symbols for the orthorhombic space groups (space-group number placed between parentheses), listed according to point group as indicated in the headline*

Number of distinct projections	222	mm2	2/m2/m2/m
6 (22 space groups)		$Pmc2_1$ (26)	$P\,2_1/m\,2/m\,2/a$ (51)
		$Pma2$ (28)	$P\,2/n\,2_1/n\,2/a$ (52)
		$Pca2_1$ (29)	$P\,2/m\,2/n\,2_1/a$ (53)
		$Pnc2$ (30)	$P\,2_1/c\,2/c\,2/a$ (54)
		$Pmn2_1$ (31)	$P\,2/b\,2_1/c\,2_1/m$ (57)
		$Pna2_1$ (33)	$P\,2_1/b\,2/c\,2_1/n$ (60)
		$Cmc2_1$ (36)	$P\,2_1/n\,2_1/m\,2_1/a$ (62)
		$Amm2$ (38)	$C\,2/m\,2/c\,2_1/m$ (63)
		$Abm2$ (39)	$C\,2/m\,2/c\,2_1/a$ (64)
		$Ama2$ (40)	$I\,2_1/m\,2_1/m\,2_1/a$ (74)
		$Aba2$ (41)	
		$Ima2$ (46)	
3 (25 space groups)	$P222_1$ (17)	$Pmm2$ (25)	$P\,2/c\,2/c\,2/m$ (49)
	$P2_12_12$ (18)	$Pcc2$ (27)	$P\,2/b\,2/a\,2/n$ (50)
	$C222_1$ (20)	$Pba2$ (32)	$P\,2_1/b\,2_1/a\,2/m$ (55)
	$C222$ (21)	$Pnn2$ (34)	$P\,2_1/c\,2_1/c\,2/n$ (56)
		$Cmm2$ (35)	$P\,2_1/n\,2_1/n\,2/m$ (58)
		$Ccc2$ (37)	$P\,2_1/m\,2_1/m\,2/n$ (59)
		$Fmm2$ (42)	$C\,2/m\,2/m\,2/m$ (65)
		$Fdd2$ (43)	$C\,2/c\,2/c\,2/m$ (66)
		$Imm2$ (44)	$C\,2/m\,2/m\,2/a$ (67)
		$Iba2$ (45)	$C\,2/c\,2/c\,2/a$ (68)
			$I\,2/b\,2/a\,2/m$ (72)
2 (2 space groups)			$P\,2_1/b\,2_1/c\,2_1/a$ (61)
			$I\,2_1/b\,2_1/c\,2_1/a$ (73)
1 (10 space groups)	$P222$ (16)		$P\,2/m\,2/m\,2/m$ (47)
	$P2_12_12_1$ (19)		$P\,2/n\,2/n\,2/n$ (48)
	$F222$ (22)		$F\,2/m\,2/m\,2/m$ (69)
	$I222$ (23)		$F\,2/d\,2/d\,2/d$ (70)
	$I2_12_12_1$ (24)		$I\,2/m\,2/m\,2/m$ (71)
Total: (59)	(9)	(22)	(28)

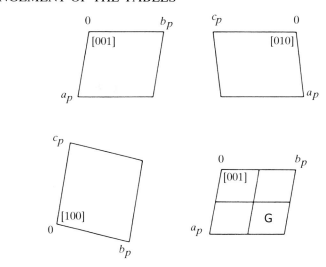

Fig. 2.2.6.1. Triclinic space groups (G = general-position diagram).

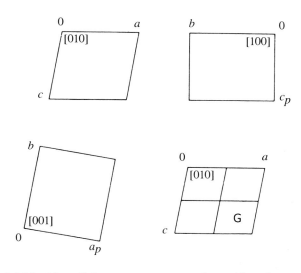

Fig. 2.2.6.2. Monoclinic space groups, setting with unique axis b (G = general-position diagram).

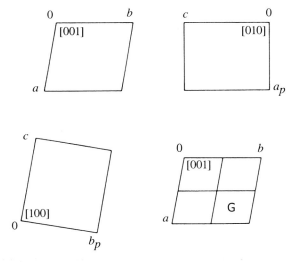

Fig. 2.2.6.3. Monoclinic space groups, setting with unique axis c.

(lower right). The projected axes, the origins and the projection directions of these diagrams are illustrated in Fig. 2.2.6.5. They refer to the so-called 'standard setting' of the space group, *i.e.* the setting described in the space-group tables and indicated by the 'standard Hermann–Mauguin symbol' in the headline.

For each orthorhombic space group, *six settings* exist, *i.e.* six different ways of assigning the labels a, b, c to the three orthorhombic symmetry directions; thus the shape and orientation of the cell are the same for each setting. These settings correspond to the six permutations of the labels of the axes (including the identity permutation); *cf.* Section 2.2.16:

$$\text{abc} \quad \text{ba}\bar{\text{c}} \quad \text{cab} \quad \bar{\text{c}}\text{ba} \quad \text{bca} \quad \text{a}\bar{\text{c}}\text{b}.$$

The symbol for each setting, here called 'setting symbol', is a short-hand notation for the transformation of the basis vectors of the

21

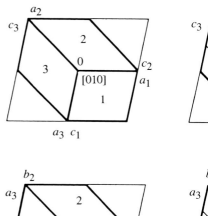
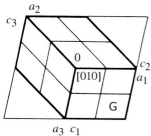
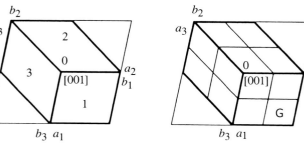

Fig. 2.2.6.4. Monoclinic space groups, cell choices 1, 2, 3. Upper pair of diagrams: setting with unique axis b. Lower pair of diagrams: setting with unique axis c. The numbers 1, 2, 3 within the cells and the subscripts of the labels of the axes indicate the cell choice (*cf.* Section 2.2.16). The unique axis points upwards from the page.

standard setting, **a, b, c**, into those of the setting considered. For instance, the setting symbol **cab** stands for the cyclic permutation

$$\mathbf{a'} = \mathbf{c}, \quad \mathbf{b'} = \mathbf{a}, \quad \mathbf{c'} = \mathbf{b}$$

or

$$(\mathbf{a'b'c'}) = (\mathbf{abc}) \begin{pmatrix} 0 & 1 & 0 \\ 0 & 0 & 1 \\ 1 & 0 & 0 \end{pmatrix} = (\mathbf{cab}),$$

where $\mathbf{a'}, \mathbf{b'}, \mathbf{c'}$ is the new set of basis vectors. An interchange of two axes reverses the handedness of the coordinate system; in order to keep the system right-handed, each interchange is accompanied by the reversal of the sense of one axis, *i.e.* by an element $\bar{1}$ in the transformation matrix. Thus, **ba$\bar{\mathbf{c}}$** denotes the transformation

$$(\mathbf{a'b'c'}) = (\mathbf{abc}) \begin{pmatrix} 0 & 1 & 0 \\ 1 & 0 & 0 \\ 0 & 0 & \bar{1} \end{pmatrix} = (\mathbf{ba\bar{c}}).$$

The six orthorhombic settings correspond to six Hermann–Mauguin symbols which, however, need not all be different; *cf.* Table 2.2.6.1.*

In the earlier (1935 and 1952) editions of *International Tables*, only one setting was illustrated, in a projection along c, so that it was usual to consider it as the 'standard setting' and to accept its cell edges as crystal axes and its space-group symbol as 'standard Hermann–Mauguin symbol'. In the present edition, however, *all six* orthorhombic settings are illustrated, as explained below.

The three projections of the symmetry elements can be interpreted in two ways. First, in the sense indicated above, that is, as different projections of a *single* (standard) setting of the space group, with the projected basis vectors **a, b, c** labelled as in Fig. 2.2.6.5. Second, each one of the three diagrams can be considered as the projection along $\mathbf{c'}$ of either one of *two different* settings: one setting in which $\mathbf{b'}$ is horizontal and one in which $\mathbf{b'}$ is vertical ($\mathbf{a'}, \mathbf{b'}, \mathbf{c'}$ refer to the setting under consideration). This second interpretation is used to illustrate in the same figure the space-group symbols corresponding to these two settings. In order to view these projections in conventional orientation ($\mathbf{b'}$ horizontal, $\mathbf{a'}$ vertical, origin in the upper left corner, projection down the positive $\mathbf{c'}$ axis), the setting with $\mathbf{b'}$ horizontal can be inspected directly with the figure upright; hence, the corresponding space-group symbol is printed above the projection. The other setting with $\mathbf{b'}$ vertical and $\mathbf{a'}$ horizontal, however, requires turning the figure over 90°, or looking at it from the side; thus, the space-group symbol is printed at the left, and it runs upwards.

The 'setting symbols' for the six settings are attached to the three diagrams of Fig. 2.2.6.6, which correspond to those of Fig. 2.2.6.5. In the orientation of the diagram where the setting symbol is read in the usual way, $\mathbf{a'}$ is vertical pointing downwards, $\mathbf{b'}$ is horizontal pointing to the right, and $\mathbf{c'}$ is pointing upwards from the page. Each setting symbol is printed in the position that in the space-group tables is actually occupied by the corresponding full Hermann–Mauguin symbol. The changes in the space-group symbol that are

---

* A space-group symbol is invariant under sign changes of the axes; *i.e.* the same symbol applies to the right-handed coordinate systems **abc, ab̄c̄, āb̄c, āb̄c̄** and the left-handed systems **ābc, āb̄c̄, ab̄c, āb̄c̄**.

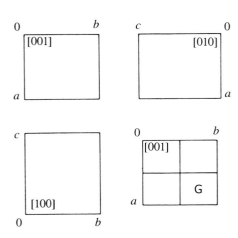

Fig. 2.2.6.5. Orthorhombic space groups. Diagrams for the 'standard setting' as described in the space-group tables (G = general-position diagram).

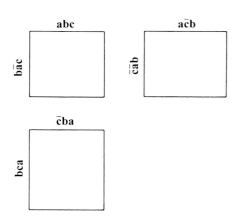

Fig. 2.2.6.6. Orthorhombic space groups. The three projections of the symmetry elements with the six setting symbols (see text). For setting symbols printed vertically, the page has to be turned clockwise by 90° or viewed from the side. Note that in the actual space-group tables instead of the setting symbols the corresponding full Hermann–Mauguin space-group symbols are printed.

(i) A symbol denoting the *type* of the symmetry operation (*cf.* Chapter 1.3), including its glide or screw part, if present. In most cases, the glide or screw part is given explicitly by fractional coordinates between parentheses. The sense of a rotation is indicated by the superscript $+$ or $-$. Abbreviated notations are used for the glide reflections $a(\frac{1}{2}, 0, 0) \equiv a$; $b(0, \frac{1}{2}, 0) \equiv b$; $c(0, 0, \frac{1}{2}) \equiv c$. Glide reflections with complicated and unconventional glide parts are designated by the letter g, followed by the glide part between parentheses.

(ii) A coordinate triplet indicating the *location* and *orientation* of the symmetry element which corresponds to the symmetry operation. For rotoinversions, the location of the inversion point is given in addition.

Details of this symbolism are presented in Section 11.1.2.

Examples
(1) a $x, y, \frac{1}{4}$
   Glide reflection with glide component $(\frac{1}{2}, 0, 0)$ through the plane $x, y, \frac{1}{4}$, *i.e.* the plane parallel to (001) at $z = \frac{1}{4}$.
(2) $\bar{4}^+$ $\frac{1}{4}, \frac{1}{4}, z$; $\frac{1}{4}, \frac{1}{4}, \frac{1}{4}$
   Fourfold rotoinversion, consisting of a counter clockwise rotation by 90° around the line $\frac{1}{4}, \frac{1}{4}, z$, followed by an inversion through the point $\frac{1}{4}, \frac{1}{4}, \frac{1}{4}$.
(3) $g(\frac{1}{4}, \frac{1}{4}, \frac{1}{2})$ x, x, z
   Glide reflection with glide component $(\frac{1}{4}, \frac{1}{4}, \frac{1}{2})$ through the plane x, x, z, *i.e.* the plane parallel to $(1\bar{1}0)$ containing the point $0, 0, 0$.
(4) $g(\frac{1}{3}, \frac{1}{6}, \frac{1}{6})$ $2x - \frac{1}{2}, x, z$ (hexagonal axes)
   Glide reflection with glide component $(\frac{1}{3}, \frac{1}{6}, \frac{1}{6})$ through the plane $2x - \frac{1}{2}, x, z$, *i.e.* the plane parallel to $(1\bar{2}10)$, which intersects the a axis at $-\frac{1}{2}$ and the b axis at $\frac{1}{4}$; this operation occurs in $R\bar{3}c$ (167, hexagonal axes).
(5) Symmetry operations in *Ibca* (73)
   Under the subheading 'For (0, 0, 0)+ set', the operation generating the coordinate triplet (2) $\bar{x} + \frac{1}{2}, \bar{y}, z + \frac{1}{2}$ from (1) x, y, z is symbolized by $2(0, 0, \frac{1}{2})$ $\frac{1}{4}, 0, z$. This indicates a twofold screw rotation with screw part $(0, 0, \frac{1}{2})$ for which the corresponding screw axis coincides with the line $\frac{1}{4}, 0, z$, *i.e.* runs parallel to [001] through the point $\frac{1}{4}, 0, 0$. Under the subheading 'For $(\frac{1}{2}, \frac{1}{2}, \frac{1}{2})$+ set', the operation generating the coordinate triplet (2) $\bar{x}, \bar{y} + \frac{1}{2}, z$ from (1) x, y, z is symbolized by 2 $0, \frac{1}{4}, z$. It is thus a twofold rotation (without screw part) around the line $0, \frac{1}{4}, z$.

2.2.10. Generators

The line *Generators selected* states the symmetry operations and their sequence, selected to generate all symmetrically equivalent points of the *General position* from a point with coordinates x, y, z. Generating translations are listed as $t(1, 0, 0)$, $t(0, 1, 0)$, $t(0, 0, 1)$; likewise for additional centring translations. The other symmetry operations are given as numbers (p) that refer to the corresponding coordinate triplets of the general position and the corresponding entries under *Symmetry operations*, as explained in Section 2.2.9 [for centred space groups the first block 'For (0, 0, 0)+ set' must be used].

For all space groups, the identity operation given by (1) is selected as the first generator. It is followed by the generators $t(1, 0, 0)$, $t(0, 1, 0)$, $t(0, 0, 1)$ of the integral lattice translations and, if necessary, by those of the centring translations, *e.g.* $t(\frac{1}{2}, \frac{1}{2}, 0)$ for a C lattice. In this way, point x, y, z and all its translationally equivalent points are generated. (The remark 'and its translationally equivalent points' will hereafter be omitted.) The sequence chosen

for the generators following the translations depends on the crystal class of the space group and is set out in Table 8.3.5.1 of Section 8.3.5.

Example: $P12_1/c1$ (14, unique axis b, cell choice 1)
   After the generation of (1) x, y, z, the operation (2) which stands for a twofold screw rotation around the axis $0, y, \frac{1}{4}$ generates point (2) of the general position with coordinate triplet $\bar{x}, y + \frac{1}{2}, \bar{z} + \frac{1}{2}$. Finally, the inversion (3) generates point (3) $\bar{x}, \bar{y}, \bar{z}$ from point (1), and point (4') $x, \bar{y} - \frac{1}{2}, z - \frac{1}{2}$ from point (2). Instead of (4'), however, the coordinate triplet (4) $x, \bar{y} + \frac{1}{2}, z + \frac{1}{2}$ is listed, because the coordinates are reduced modulo 1.

The example shows that for the space group $P12_1/c1$ two operations, apart from the identity and the generating translations, are sufficient to generate all symmetrically equivalent points. Alternatively, the inversion (3) plus the glide reflection (4), or the glide reflection (4) plus the twofold screw rotation (2), might have been chosen as generators. The process of generation and the selection of the generators for the space-group tables, as well as the resulting sequence of the symmetry operations, are discussed in Section 8.3.5.

For different descriptions of the same space group (settings, cell choices, origin choices), the generating operations are the same. Thus, the transformation relating the two coordinate systems transforms also the generators of one description into those of the other.

From the Fifth Edition onwards, this applies also to the description of the seven rhombohedral (R) space groups by means of 'hexagonal' and 'rhombohedral' axes. In previous editions, there was a difference in the *sequence* (not the data) of the 'coordinate triplets' and the 'symmetry operations' in both descriptions (*cf.* Section 2.10 in the First to Fourth Editions).

2.2.11. Positions

The entries under *Positions** (more explicitly called *Wyckoff positions*) consist of the one *General position* (upper block) and the *Special positions* (blocks below). The columns in each block, from left to right, contain the following information for each Wyckoff position.

(i) *Multiplicity M of the Wyckoff position.* This is the number of equivalent points per unit cell. For primitive cells, the multiplicity M of the general position is equal to the order of the point group of the space group; for centred cells, M is the product of the order of the point group and the number (2, 3 or 4) of lattice points per cell. The multiplicity of a special position is always a divisor of the multiplicity of the general position.

(ii) *Wyckoff letter.* This letter is merely a coding scheme for the Wyckoff positions, starting with a at the bottom position and continuing upwards in alphabetical order (the theoretical background on Wyckoff positions is given in Section 8.3.2).

(iii) *Site symmetry.* This is explained in Section 2.2.12.

(iv) *Coordinates.* The sequence of the coordinate triplets is based on the *Generators* (*cf.* Section 2.2.10). For centred space groups, the centring translations, for instance $(0, 0, 0)+$ $(\frac{1}{2}, \frac{1}{2}, \frac{1}{2})+$, are listed above the coordinate triplets. The symbol '$+$' indicates that, in order to obtain a complete Wyckoff position, the components of

---

* The term *Position* (singular) is defined as a *set* of symmetrically equivalent points, in agreement with *IT* (1935): Point position; *Punktlage* (German); *Position* (French). Note that in *IT* (1952) the plural, equivalent positions, was used.

these centring translations have to be added to the listed coordinate triplets. Note that not all points of a position always lie within the unit cell; some may be outside since the coordinates are formulated modulo 1; thus, for example, $\bar{x}, \bar{y}, \bar{z}$ is written rather than $\bar{x}+1, \bar{y}+1, \bar{z}+1$.

The M coordinate triplets of a position represent the coordinates of the M equivalent points (atoms) in the unit cell. A graphic representation of the points of the general position is provided by the general-position diagram; cf. Section 2.2.6.

(v) *Reflection conditions.* These are described in Section 2.2.13.

The two types of positions, general and special, are characterized as follows:

(i) *General position*
A set of symmetrically equivalent points, *i.e.* a 'crystallographic orbit', is said to be in 'general position' if each of its points is left invariant only by the identity operation but by no other symmetry operation of the space group. Each space group has only one general position.

The coordinate triplets of a general position (which always start with *x*, *y*, *z*) can also be interpreted as a short-hand form of the matrix representation of the symmetry operations of the space group; this viewpoint is further described in Sections 8.1.6 and 11.1.1.

(ii) *Special position(s)*
A set of symmetrically equivalent points is said to be in 'special position' if each of its points is mapped onto itself by the identity and at least one further symmetry operation of the space group. This implies that specific constraints are imposed on the coordinates of each point of a special position; *e.g.* $x = \frac{1}{4}, y = 0$, leading to the triplet $\frac{1}{4}, 0, z$; or $y = x + \frac{1}{2}$, leading to the triplet $x, x + \frac{1}{2}, z$. The number of special positions in a space group [up to 26 in *Pmmm* (No. 47)] depends on the number and types of symmetry operations that map a point onto itself.

The set of *all* symmetry operations that map a point onto itself forms a group, known as the 'site-symmetry group' of that point. It is given in the third column by the 'oriented site-symmetry symbol' which is explained in Section 2.2.12. General positions always have site symmetry 1, whereas special positions have higher site symmetries, which can differ from one special position to another.

If in a crystal structure the centres of finite objects, such as molecules, are placed at the points of a special position, each such object must display a point symmetry that is at least as high as the site symmetry of the special position. Geometrically, this means that the centres of these objects are located on symmetry elements without translations (centre of symmetry, mirror plane, rotation axis, rotoinversion axis) or at the intersection of several symmetry elements of this kind (*cf.* space-group diagrams).

Note that the location of an object on a screw axis or on a glide plane does *not* lead to an increase in the site symmetry and to a consequent reduction of the multiplicity for that object. Accordingly, a space group that contains only symmetry elements *with* translation components does not have any special position. Such a space group is called 'fixed-point-free'. The 13 space groups of this kind are listed in Section 8.3.2.

Example: Space group $C12/c1$ (15, unique axis *b*, cell choice 1)
The general position $8f$ of this space group contains eight equivalent points per cell, each with site symmetry 1. The coordinate triplets of four points, (1) to (4), are given explicitly, the coordinates of the other four points are obtained by adding the components $\frac{1}{2}, \frac{1}{2}, 0$ of the C-centring translation to the coordinate triplets (1) to (4).

The space group has five special positions with Wyckoff letters *a* to *e*. The positions 4*a* to 4*d* require inversion symmetry, $\bar{1}$, whereas Wyckoff position 4*e* requires twofold rotation symmetry, 2, for any object in such a position. For position 4*e*, for instance, the four equivalent points have the coordinates $0, y, \frac{1}{4}$; $0, \bar{y}, \frac{3}{4}$; $\frac{1}{2}, y + \frac{1}{2}, \frac{1}{4}$; $\frac{1}{2}, \bar{y} + \frac{1}{2}, \frac{3}{4}$. The values of *x* and *z* are specified, whereas *y* may take any value. Since each point of position 4*e* is mapped onto itself by a twofold rotation, the multiplicity of the position is reduced from 8 to 4, whereas the order of the site-symmetry group is increased from 1 to 2.

From the entries 'Symmetry operations', the locations of the four twofold axes can be deduced as $0, y, \frac{1}{4}$; $0, y, \frac{3}{4}$; $\frac{1}{2}, y, \frac{1}{4}$; $\frac{1}{2}, y, \frac{3}{4}$.

From this example, the general rule is apparent that the product of the position multiplicity and the order of the corresponding site-symmetry group is constant for all Wyckoff positions of a given space group; it is the multiplicity of the general position.

Attention is drawn to ambiguities in the description of crystal structures in a few space groups, depending on whether the coordinate triplets of *IT* (1952) or of this edition are taken. This problem is analysed by Parthé *et al.* (1988).

2.2.12. Oriented site-symmetry symbols

The third column of each Wyckoff position gives the *Site symmetry** of that position. The site-symmetry group is isomorphic to a (proper or improper) subgroup of the point group to which the space group under consideration belongs. The site-symmetry groups of the different points of the same special position are conjugate (symmetrically equivalent) subgroups of the space group. For this reason, all points of one special position are described by the same site-symmetry symbol.

Oriented site-symmetry symbols (*cf.* Fischer *et al.*, 1973) are employed to show how the symmetry elements at a site are related to the symmetry elements of the crystal lattice. The site-symmetry symbols display the same sequence of symmetry directions as the space-group symbol (*cf.* Table 2.2.4.1). Sets of equivalent symmetry directions that do not contribute any element to the site-symmetry group are represented by a dot. In this way, the orientation of the symmetry elements at the site is emphasized, as illustrated by the following examples.

Examples
(1) In the tetragonal space group $P4_22_12$ (94), Wyckoff position 4*f* has site symmetry ..2 and position 2*b* has site symmetry 2.22. The easiest way to interpret the symbols is to look at the dots first. For position 4*f*, the 2 is preceded by two dots and thus must belong to a tertiary symmetry direction. Only one tertiary direction is used. Consequently, the site symmetry is the monoclinic point group 2 with one of the two tetragonal tertiary directions as twofold axis.

Position *b* has one dot, with one symmetry symbol before and two symmetry symbols after it. The dot corresponds, therefore, to the secondary symmetry directions. The first symbol 2 indicates a twofold axis along the primary symmetry direction (*c* axis). The final symbols 22 indicate two twofold axes along the two mutually perpendicular tertiary directions [$1\bar{1}0$] and [110]. The site symmetry is thus orthorhombic, 222.

(2) In the cubic space group $I23$ (197), position 6*b* has 222.. as its oriented site-symmetry symbol. The orthorhombic group 222 is completely related to the primary set of cubic symmetry

---

* Often called point symmetry: *Punktsymmetrie* or *Lagesymmetrie* (German): *symétrie ponctuelle* (French).

directions, with the three twofold axes parallel to the three equivalent primary directions [100], [010], [001].

(3) In the cubic space group $Pn\bar{3}n$ (222), position 6b has 42.2 as its site-symmetry symbol. This 'cubic' site-symmetry symbol displays a tetragonal site symmetry. The position of the dot indicates that there is no symmetry along the four secondary cubic directions. The fourfold axis is connected with one of the three primary cubic symmetry directions and two equivalent twofold axes occur along the remaining two primary directions. Moreover, the group contains two mutually perpendicular (equivalent) twofold axes along those two of the six tertiary cubic directions $\langle 110 \rangle$ that are normal to the fourfold axis. Each pair of equivalent twofold axes is given by just one symbol 2. (Note that at the six sites of position 6b the fourfold axes are twice oriented along a, twice along b and twice along c.)

(4) In the tetragonal space group $P4_2/nnm$ (134), position 2a has site symmetry $\bar{4}2m$. The site has symmetry for all symmetry directions. Because of the presence of the primary $\bar{4}$ axis, only one of the twofold axes along the two secondary directions need be given explicitly and similarly for the mirror planes m perpendicular to the two tertiary directions.

The above examples show:

(i) The oriented site-symmetry symbols become identical to Hermann–Mauguin point-group symbols if the dots are omitted.

(ii) Sets of symmetry directions having more than one equivalent direction may require more than one character if the site-symmetry group belongs to a lower crystal system than the space group under consideration.

To show, for the same type of site symmetry, how the oriented site-symmetry symbol depends on the space group under discussion, the site-symmetry group mm2 will be considered in orthorhombic and tetragonal space groups. Relevant crystal classes are mm2, mmm, 4mm, $\bar{4}2m$ and 4/mmm. The site symmetry mm2 contains two mutually perpendicular mirror planes intersecting in a twofold axis.

For space groups of crystal class mm2, the twofold axis at the site must be parallel to the one direction of the rotation axes of the space group. The site-symmetry group mm2, therefore, occurs only in the orientation mm2. For space groups of class mmm (full symbol 2/m 2/m 2/m), the twofold axis at the site may be parallel to a, b or c and the possible orientations of the site symmetry are 2mm, m2m and mm2. For space groups of the tetragonal crystal class 4mm, the twofold axis of the site-symmetry group mm2 must be parallel to the fourfold axis of the crystal. The two mirror planes must belong either to the two secondary or to the two tertiary tetragonal directions so that 2mm. and 2.mm are possible site-symmetry symbols. Similar considerations apply to class $\bar{4}2m$ which can occur in two settings, $\bar{4}2m$ and $\bar{4}m2$. Finally, for class 4/mmm (full symbol 4/m 2/m 2/m), the twofold axis of 2mm may belong to any of the three kinds of symmetry directions and possible oriented site symmetries are 2mm., 2.mm, m2m. and m.2m. In the first two symbols, the twofold axis extends along the single primary direction and the mirror planes occupy either both secondary or both tertiary directions; in the last two cases, one mirror plane belongs to the primary direction and the second to either one secondary or one tertiary direction (the other equivalent direction in each case being occupied by the twofold axis).

---

* The reflection conditions were called *Auslöschungen* (German), missing spectra (English) and *extinctions* (French) in *IT* (1935) and 'Conditions limiting possible reflections' in *IT* (1952); they are often referred to as 'Systematic or space-group absences' (*cf.* Chapter 12.3).

Table 2.2.13.1. *Integral reflection conditions for centred cells (lattices)*

Reflection condition	Centring type of cell	Centring symbol
None	Primitive	$\begin{cases} P \\ R\,\text{*(rhombohedral axes)} \end{cases}$
$h + k = 2n$	C-face centred	C
$k + l = 2n$	A-face centred	A
$h + l = 2n$	B-face centred	B
$h + k + l = 2n$	Body centred	I
$h + k, h + l$ and $k + l = 2n$ or: h, k, l all odd or all even ('unmixed')	All-face centred	F
$-h + k + l = 3n$	Rhombohedrally centred, obverse setting (standard)	$\left.\begin{array}{c} \\ \\ \\ \\ \\ \end{array}\right\}R\,\text{* (hexagonal axes)}$
$h - k + l = 3n$	Rhombohedrally centred, reverse setting	
$h - k = 3n$	Hexagonally centred	H†

* For further explanations see Chapters 1.2 and 2.1.
† For the use of the unconventional H cell, see Chapter 1.2.

2.2.13. Reflection conditions

The *Reflection conditions** are listed in the right-hand column of each Wyckoff position.

These conditions are formulated here, in accordance with general practice, as 'conditions of occurrence' (structure factor not systematically zero) and not as 'extinctions' or 'systematic absences' (structure factor zero). Reflection conditions are listed for *all* those three-, two- and one-dimensional sets of reflections for which extinctions exist; hence, for those nets or rows that are *not* listed, no reflection conditions apply.

There are two types of systematic reflection conditions for diffraction of crystals by radiation:

(1) *General conditions.* They apply to *all* Wyckoff positions of a space group, *i.e.* they are always obeyed, irrespective of which Wyckoff positions are occupied by atoms in a particular crystal structure.

(2) *Special conditions* ('extra' conditions). They apply only to *special* Wyckoff positions and occur always in addition to the general conditions of the space group. Note that each extra condition is valid only for the scattering contribution of those atoms that are located in the relevant special Wyckoff position. If the special position is occupied by atoms whose scattering power is high, in comparison with the other atoms in the structure, reflections violating the extra condition will be weak.

2.2.13.1. *General reflection conditions*

These are due to one of three effects:

(i) *Centred cells.* The resulting conditions apply to the whole three-dimensional set of reflections hkl. Accordingly, they are called *integral reflection conditions.* They are given in Table 2.2.13.1. These conditions result from the centring vectors of centred cells. They disappear if a primitive cell is chosen instead of a centred cell. Note that the centring symbol and the corresponding integral reflection condition may change with a change of the basis vectors (*e.g.* monoclinic: $C \rightarrow A \rightarrow I$).

Table 2.2.13.2. *Zonal and serial reflection conditions for glide planes and screw axes (cf. Chapter 1.3)*

(a) Glide planes

| Type of reflections | Reflection condition | Glide plane | | | Crystallographic coordinate system to which condition applies |
		Orientation of plane	Glide vector	Symbol	
$0kl$	$k = 2n$	(100)	$\mathbf{b}/2$	b	Monoclinic (a unique), Tetragonal; Orthorhombic, Cubic
	$l = 2n$		$\mathbf{c}/2$	c	
	$k + l = 2n$		$\mathbf{b}/2 + \mathbf{c}/2$	n	
	$k + l = 4n$ $(k, l = 2n)^*$		$\mathbf{b}/4 \pm \mathbf{c}/4$	d	
$h0l$	$l = 2n$	(010)	$\mathbf{c}/2$	c	Monoclinic (b unique), Tetragonal; Orthorhombic, Cubic
	$h = 2n$		$\mathbf{a}/2$	a	
	$l + h = 2n$		$\mathbf{c}/2 + \mathbf{a}/2$	n	
	$l + h = 4n$ $(l, h = 2n)^*$		$\mathbf{c}/4 \pm \mathbf{a}/4$	d	
$hk0$	$h = 2n$	(001)	$\mathbf{a}/2$	a	Monoclinic (c unique), Tetragonal; Orthorhombic, Cubic
	$k = 2n$		$\mathbf{b}/2$	b	
	$h + k = 2n$		$\mathbf{a}/2 + \mathbf{b}/2$	n	
	$h + k = 4n$ $(h, k = 2n)^*$		$\mathbf{a}/4 \pm \mathbf{b}/4$	d	
$h\bar{h}0l$ $0k\bar{k}l$ $\bar{h}0hl$	$l = 2n$	(11$\bar{2}$0) ($\bar{2}$110) (1$\bar{2}$10) $\}\{11\bar{2}0\}$	$\mathbf{c}/2$	c	Hexagonal
$hh.\overline{2h}.l$ $\overline{2h}.hhl$ $h.\overline{2h}.hl$	$l = 2n$	(1$\bar{1}$00) (01$\bar{1}$0) ($\bar{1}$010) $\}\{1\bar{1}00\}$	$\mathbf{c}/2$	c	Hexagonal
hhl hkk hkh	$l = 2n$ $h = 2n$ $k = 2n$	(1$\bar{1}$0) (01$\bar{1}$) ($\bar{1}$01) $\}\{1\bar{1}0\}$	$\mathbf{c}/2$ $\mathbf{a}/2$ $\mathbf{b}/2$	c, n a, n b, n	Rhombohedral†
$hhl, h\bar{h}l$	$l = 2n$	(1$\bar{1}$0), (110)	$\mathbf{c}/2$	c, n	Tetragonal‡
	$2h + l = 4n$		$\mathbf{a}/4 \pm \mathbf{b}/4 \pm \mathbf{c}/4$	d	
$hkk, hk\bar{k}$	$h = 2n$	(01$\bar{1}$), (011)	$\mathbf{a}/2$	a, n	Cubic§
	$2k + h = 4n$		$\pm\mathbf{a}/4 + \mathbf{b}/4 \pm \mathbf{c}/4$	d	
$hkh, \bar{h}kh$	$k = 2n$	($\bar{1}$01), (101)	$\mathbf{b}/2$	b, n	
	$2h + k = 4n$		$\pm\mathbf{a}/4 \pm \mathbf{b}/4 + \mathbf{c}/4$	d	

* Glide planes d with orientations (100), (010) and (001) occur only in orthorhombic and cubic F space groups. Combination of the integral reflection condition (hkl: all odd or all even) with the zonal conditions for the d glide planes leads to the further conditions given between parentheses.

† For rhombohedral space groups described with 'rhombohedral axes' the three reflection conditions ($l = 2n, h = 2n, k = 2n$) imply interleaving of c and n glides, a and n glides, b and n glides, respectively. In the Hermann–Mauguin space-group symbols, c is always used, as in $R3c$ (161) and $R\bar{3}c$ (167), because c glides occur also in the hexagonal description of these space groups.

‡ For tetragonal P space groups, the two reflection conditions (hhl and $h\bar{h}l$ with $l = 2n$) imply interleaving of c and n glides. In the Hermann–Mauguin space-group symbols, c is always used, irrespective of which glide planes contain the origin: cf. $P4cc$ (103), $P\bar{4}2c$ (112) and $P4/nnc$ (126).

§ For cubic space groups, the three reflection conditions ($l = 2n, h = 2n, k = 2n$) imply interleaving of c and n glides, a and n glides, and b and n glides, respectively. In the Hermann–Mauguin space-group symbols, either c or n is used, depending upon which glide plane contains the origin, cf. $P\bar{4}3n$ (218), $Pn\bar{3}n$ (222), $Pm\bar{3}n$ (223) vs $F\bar{4}3c$ (219), $Fm\bar{3}c$ (226), $Fd\bar{3}c$ (228).

(ii) *Glide planes.* The resulting conditions apply only to two-dimensional sets of reflections, *i.e.* to reciprocal-lattice nets containing the origin (such as $hk0$, $h0l$, $0kl$, hhl). For this reason, they are called *zonal reflection conditions*. The indices hkl of these 'zonal reflections' obey the relation $hu + kv + lw = 0$, where $[uvw]$, the direction of the zone axis, is normal to the reciprocal-lattice net.

Note that the symbol of a glide plane and the corresponding zonal reflection condition may change with a change of the basis vectors (*e.g.* monoclinic: $c \rightarrow n \rightarrow a$).

(iii) *Screw axes.* The resulting conditions apply only to one-dimensional sets of reflections, *i.e.* reciprocal-lattice rows contain-

Table 2.2.13.2. (*cont.*)

(*b*) Screw axes

Type of reflections	Reflection conditions	Screw axis			Crystallographic coordinate system to which condition applies
		Direction of axis	Screw vector	Symbol	
$h00$	$h = 2n$	[100]	$\mathbf{a}/2$	2_1	Monoclinic (*a* unique), Orthorhombic, Tetragonal
				4_2	Cubic
	$h = 4n$		$\mathbf{a}/4$	$4_1, 4_3$	
$0k0$	$k = 2n$	[010]	$\mathbf{b}/2$	2_1	Monoclinic (*b* unique), Orthorhombic, Tetragonal
				4_2	Cubic
	$k = 4n$		$\mathbf{b}/4$	$4_1, 4_3$	
$00l$	$l = 2n$	[001]	$\mathbf{c}/2$	2_1	Monoclinic (*c* unique), Orthorhombic
				4_2	Tetragonal, Cubic
	$l = 4n$		$\mathbf{c}/4$	$4_1, 4_3$	
$000l$	$l = 2n$	[001]	$\mathbf{c}/2$	6_3	Hexagonal
	$l = 3n$		$\mathbf{c}/3$	$3_1, 3_2, 6_2, 6_4$	
	$l = 6n$		$\mathbf{c}/6$	$6_1, 6_5$	

ing the origin (such as $h00$, $0k0$, $00l$). They are called *serial reflection conditions*.

Reflection conditions of types (ii) and (iii) are listed in Table 2.2.13.2. They can be understood as follows: Zonal and serial reflections form two- or one-dimensional sections through the origin of reciprocal space. In direct space, they correspond to projections of a crystal structure onto a plane or onto a line. Glide planes or screw axes may reduce the translation periods in these projections (*cf.* Section 2.2.14) and thus decrease the size of the projected cell. As a consequence, the cells in the corresponding reciprocal-lattice sections are increased, which means that systematic absences of reflections occur.

For the two-dimensional groups, the reasoning is analogous. The reflection conditions for the plane groups are assembled in Table 2.2.13.3.

Table 2.2.13.3. *Reflection conditions for the plane groups*

Type of reflections	Reflection condition	Centring type of plane cell; or glide line with glide vector	Coordinate system to which condition applies
hk	None	Primitive *p*	All systems
	$h + k = 2n$	Centred *c*	Rectangular
	$h - k = 3n$	Hexagonally centred *h**	Hexagonal
$h0$	$h = 2n$	Glide line *g* normal to *b* axis; glide vector $\frac{1}{2}\mathbf{a}$	Rectangular, Square
$0k$	$k = 2n$	Glide line *g* normal to *a* axis; glide vector $\frac{1}{2}\mathbf{b}$	

* For the use of the unconventional *h* cell see Chapter 1.2.

For the *interpretation of observed reflections*, the general reflection conditions must be studied in the order (i) to (iii), as conditions of type (ii) may be included in those of type (i), while conditions of type (iii) may be included in those of types (i) or (ii). This is shown in the example below.

In the *space-group tables*, the reflection conditions are given according to the following rules:

(i) for a given space group, *all* reflection conditions are listed; hence for those nets or rows that are *not* listed no conditions apply. No distinction is made between 'independent' and 'included' conditions, as was done in *IT* (1952), where 'included' conditions were placed in parentheses;

(ii) the integral condition, if present, is always listed first, followed by the zonal and serial conditions;

(iii) conditions that have to be satisified simultaneously are separated by a comma or by 'AND'. Thus, if two indices must be even, say *h* and *l*, the condition is written $h, l = 2n$ rather than $h = 2n$ and $l = 2n$. The same applies to sums of indices. Thus, there are several different ways to express the integral conditions for an *F*-centred lattice: '$h + k, h + l, k + l = 2n$' or '$h + k, h + l = 2n$ and $k + l = 2n$' or '$h + k = 2n$ and $h + l, k + l = 2n$' (*cf.* Table 2.2.13.1);

(iv) conditions separated by 'OR' are alternative conditions. For example, '$hkl : h = 2n + 1$ or $h + k + l = 4n$' means that *hkl* is 'present' if either the condition $h = 2n + 1$ *or* the alternative condition $h + k + l = 4n$ is fulfilled. Obviously, *hkl* is a 'present' reflection also if both conditions are satisfied. Note that 'or' conditions occur only for the *special conditions* described in Section 2.2.13.2;

(v) in crystal systems with two or more symmetrically equivalent nets or rows (tetragonal and higher), only *one* representative set (the first one in Table 2.2.13.2) is listed; *e.g.* tetragonal: only the first members of the equivalent sets $0kl$ and $h0l$ or $h00$ and $0k0$ are listed;

(vi) for cubic space groups, it is stated that the indices *hkl* are 'cyclically permutable' or 'permutable'. The cyclic permutability of

h, k and l in all rhombohedral space groups, described with 'rhombohedral axes', and of h and k in some tetragonal space groups are not stated;

(vii) in the 'hexagonal-axes' descriptions of trigonal and hexagonal space groups, Bravais–Miller indices $hkil$ are used. They obey two conditions:

(a) $h + k + i = 0$, i.e. $i = -(h + k)$;

(b) the indices h, k, i are cyclically permutable; this is not stated. Further details can be found in textbooks of crystallography.

Note that the integral reflection conditions for a rhombohedral lattice, described with 'hexagonal axes', permit the presence of only one member of the pair $hkil$ and $\bar{h}\bar{k}\bar{i}l$ for $l \neq 3n$ (cf. Table 2.2.13.1). This applies also to the zonal reflections $hh0l$ and $h\bar{h}0l$, which for the rhombohedral space groups must be considered separately.

Example

For a monoclinic crystal (b unique), the following reflection conditions have been observed:

(1) hkl: $h + k = 2n$;

(2) $0kl$: $k = 2n$; $h0l$: $h, l = 2n$; $hk0$: $h + k = 2n$;

(3) $h00$: $h = 2n$; $0k0$: $k = 2n$; $00l$: $l = 2n$.

Line (1) states that the cell used for the description of the space group is C centred. In line (2), the conditions $0kl$ with $k = 2n$, $h0l$ with $h = 2n$ and $hk0$ with $h + k = 2n$ are a consequence of the integral condition (1), leaving only $h0l$ with $l = 2n$ as a new condition. This indicates a glide plane c. Line (3) presents no new condition, since $h00$ with $h = 2n$ and $0k0$ with $k = 2n$ follow from the integral condition (1), whereas $00l$ with $l = 2n$ is a consequence of a zonal condition (2). Accordingly, there need not be a twofold screw axis along [010]. Space groups obeying the conditions are Cc (9, b unique, cell choice 1) and $C2/c$ (15, b unique, cell choice 1). On the basis of diffraction symmetry and reflection conditions, no choice between the two space groups can be made (cf. Part 3).

For a different choice of the basis vectors, the reflection conditions would appear in a different form owing to the transformation of the reflection indices (cf. cell choices 2 and 3 for space groups Cc and $C2/c$ in Part 7).

2.2.13.2. Special or 'extra' reflection conditions

These apply either to the integral reflections hkl or to particular sets of zonal or serial reflections. In the space-group tables, the minimal special conditions are listed that, on combination with the general conditions, are sufficient to generate the complete set of conditions. This will be apparent from the examples below.

Examples

(1) $P4_22$ (93)

General position $8p$: $00l$: $l = 2n$, due to 4_2; the projection on [001] of any crystal structure with this space group has periodicity $\frac{1}{2}c$.

Special position $4i$: hkl: $h + k + l = 2n$; any set of symmetrically equivalent atoms in this position displays additional I centring.

Special position $4n$: $0kl$: $l = 2n$; any set of equivalent atoms in this position displays a glide plane $c \perp [100]$. Projection of this set along [100] results in a halving of the original c axis, whence the special condition. Analogously for $h0l$: $l = 2n$.

(2) $C12/c1$ (15, unique axis b, cell choice 1)

General position $8f$: hkl: $h + k = 2n$, due to the C-centred cell.

Special position $4d$: hkl: $k + l = 2n$, due to additional A and B centring for atoms in this position. Combination with the general condition results in hkl: $h + k, h + l, k + l = 2n$ or hkl all odd or all even; this corresponds to an F-centred arrangement of atoms in this position.

Special position $4b$: hkl: $l = 2n$, due to additional halving of the c axis for atoms in this position. Combination with the general condition results in hkl: $h + k, l = 2n$; this corresponds to a C-centred arrangement in a cell with half the original c axis. No further condition results from the combination.

(3) $I12/a1$ (15, unique axis b, cell choice 3)

For the description of space group No. 15 with cell choice 3 (see Section 2.2.16 and space-group tables), the reflection conditions appear as follows:

General position $8f$: hkl: $h + k + l = 2n$, due to the I-centred cell.

Special position $4b$: hkl: $h = 2n$, due to additional halving of the a axis. Combination gives hkl: $h, k + l = 2n$, i.e. an A-centred arrangement of atoms in a cell with half the original a axis.

An analogous result is obtained for position $4d$.

(4) $Fmm2$ (42)

General position $16e$: hkl: $h + k, h + l, k + l = 2n$, due to the F-centred cell.

Special position $8b$: hkl: $h = 2n$, due to additional halving of the a axis. Combination results in hkl: $h, k, l = 2n$, i.e. all indices even; the atoms in this position are arranged in a primitive lattice with axes $\frac{1}{2}a$, $\frac{1}{2}b$ and $\frac{1}{2}c$.

For the cases where the special reflection conditions are described by means of combinations of 'OR' and 'AND' instructions, the 'AND' condition always has to be evaluated with priority, as shown by the following example.

Example: $P\bar{4}3n$ (218)

Special position $6d$: hkl: $h + k + l = 2n$ or $h = 2n + 1$, $k = 4n$ and $l = 4n + 2$.

This expression contains the following two conditions:

(a) hkl: $h + k + l = 2n$;

(b) $h = 2n + 1$ and $k = 4n$ and $l = 4n + 2$.

A reflection is 'present' (occurring) if either condition (a) is satisfied or if a permutation of the three conditions in (b) are simultaneously fulfilled.

2.2.13.3. Structural or non-space-group absences

Note that in addition *non-space-group absences* may occur that are not due to the symmetry of the space group (i.e. centred cells, glide planes or screw axes). Atoms in general or special positions may cause additional systematic absences if their coordinates assume special values [e.g. 'noncharacteristic orbits' (Engel *et al.*, 1984)]. Non-space-group absences may also occur for special arrangements of atoms ('false symmetry') in a crystal structure (cf. Templeton, 1956; Sadanaga *et al.*, 1978). Non-space-group absences may occur also for polytypic structures; this is briefly discussed by Ďurovič in Section 9.2.2.2.5 of *International Tables for Crystallography* (2004), Vol. C. Even though all these 'structural absences' are fortuitous and due to the special arrangements of atoms in a particular crystal structure, they have the appearance of space-group absences. Occurrence of structural absences thus may lead to an *incorrect assignment of the space group*. Accordingly, the reflection conditions in the space-group tables must be considered as a minimal set of conditions.

The use of reflection conditions and of the symmetry of reflection intensities for space-group determination is described in Part 3.

Table 2.2.14.1. *Cell parameters a', b', γ' of the two-dimensional cell in terms of cell parameters a, b, c, α, β, γ of the three-dimensional cell for the projections listed in the space-group tables of Part 7*

| Projection direction | Triclinic | Monoclinic | | Orthorhombic | Projection direction | Tetragonal |
		Unique axis b	Unique axis c			
[001]	$a' = a\sin\beta$ $b' = b\sin\alpha$ $\gamma' = 180° - \gamma^*$ †	$a' = a\sin\beta$ $b' = b$ $\gamma' = 90°$	$a' = a$ $b' = b$ $\gamma' = \gamma$	$a' = a$ $b' = b$ $\gamma' = 90°$	[001]	$a' = a$ $b' = a$ $\gamma' = 90°$
[100]	$a' = b\sin\gamma$ $b' = c\sin\beta$ $\gamma' = 180° - \alpha^*$ †	$a' = b$ $b' = c\sin\beta$ $\gamma' = 90°$	$a' = b\sin\gamma$ $b' = c$ $\gamma' = 90°$	$a' = b$ $b' = c$ $\gamma' = 90°$	[100]	$a' = a$ $b' = c$ $\gamma' = 90°$
[010]	$a' = c\sin\alpha$ $b' = \alpha\sin\gamma$ $\gamma' = 180° - \beta^*$ †	$a' = c$ $b' = a$ $\gamma' = \beta$	$a' = c$ $b' = a\sin\gamma$ $\gamma' = 90°$	$a' = c$ $b' = a$ $\gamma' = 90°$	[110]	$a' = (a/2)\sqrt{2}$ $b' = c$ $\gamma' = 90°$

Projection direction	Hexagonal	Projection direction	Rhombohedral ‡	Projection direction	Cubic
[001]	$a' = a$ $b' = a$ $\gamma' = 120°$	[111]	$a' = \dfrac{2}{\sqrt{3}}\,a\sin(\alpha/2)$ $b' = \dfrac{2}{\sqrt{3}}\,a\sin(\alpha/2)$ $\gamma' = 120°$	[001]	$a' = a$ $b' = a$ $\gamma' = 90°$
[100]	$a' = (a/2)\sqrt{3}$ $b' = c$ $\gamma' = 90°$	[1$\bar{1}$0]	$a' = a\cos(\alpha/2)$ $b' = a$ $\gamma' = \delta$ §	[111]	$a' = a\sqrt{2/3}$ $b' = a\sqrt{2/3}$ $\gamma' = 120°$
[210]	$a' = a/2$ $b' = c$ $\gamma' = 90°$	[$\bar{2}$11]	$a' = \dfrac{1}{\sqrt{3}}\,a\sqrt{1 + 2\cos\alpha}$ $b' = a\sin(\alpha/2)$ $\gamma' = 90°$	[110]	$a' = (a/2)\sqrt{2}$ $b' = a$ $\gamma' = 90°$

† $\cos\alpha^* = \dfrac{\cos\beta\cos\gamma - \cos\alpha}{\sin\beta\sin\gamma}$; $\cos\beta^* = \dfrac{\cos\gamma\cos\alpha - \cos\beta}{\sin\gamma\sin\alpha}$; $\cos\gamma^* = \dfrac{\cos\alpha\cos\beta - \cos\gamma}{\sin\alpha\sin\beta}$.

‡ The entry 'Rhombohedral' refers to the primitive rhombohedral cell with $a = b = c, \alpha = \beta = \gamma$ (*cf.* Table 2.1.2.1).

§ $\cos\delta = \dfrac{\cos\alpha}{\cos\alpha/2}$.

2.2.14. Symmetry of special projections

Projections of crystal structures are used by crystallographers in special cases. Use of so-called 'two-dimensional data' (zero-layer intensities) results in the projection of a crystal structure along the normal to the reciprocal-lattice net.

Even though the projection of a finite object along *any* direction may be useful, the projection of a *periodic* object such as a crystal structure is only sensible along a rational lattice direction (lattice row). Projection along a nonrational direction results in a constant density in at least one direction.

2.2.14.1. *Data listed in the space-group tables*

Under the heading *Symmetry of special projections*, the following data are listed for three projections of each space group; no projection data are given for the plane groups.

(i) *The projection direction.* All projections are orthogonal, *i.e.* the projection is made onto a plane normal to the projection direction. This ensures that spherical atoms appear as circles in the projection. For each space group, three projections are listed. If a lattice has three kinds of symmetry directions, the three projection directions correspond to the primary, secondary and tertiary symmetry directions of the lattice (*cf.* Table 2.2.4.1). If a lattice

contains less than three kinds of symmetry directions, as in the triclinic, monoclinic and rhombohedral cases, the additional projection direction(s) are taken along coordinate axes, *i.e.* lattice rows lacking symmetry.

The directions for which projection data are listed are as follows:

Triclinic			
Monoclinic			
(both settings)	[001]	[100]	[010]
Orthorhombic			
Tetragonal	[001]	[100]	[110]
Hexagonal	[001]	[100]	[210]
Rhombohedral	[111]	[1$\bar{1}$0]	[$\bar{2}$1$\bar{1}$]
Cubic	[001]	[111]	[110]

(ii) *The Hermann–Mauguin symbol of the plane group* resulting from the projection of the space group. If necessary, the symbols are given in oriented form; for example, plane group *pm* is expressed either as *p*1*m*1 or as *p*11*m*.

(iii) *Relations between the basis vectors* $\mathbf{a}'$, $\mathbf{b}'$ *of the plane group and the basis vectors* $\mathbf{a}$, $\mathbf{b}$, $\mathbf{c}$ *of the space group.* Each set of basis vectors refers to the conventional coordinate system of the plane group or space group, as employed in Parts 6 and 7. The basis vectors of the two-dimensional cell are always called $\mathbf{a}'$ and $\mathbf{b}'$ irrespective of which two of the basis vectors $\mathbf{a}$, $\mathbf{b}$, $\mathbf{c}$ of the three-dimensional cell are projected to form the plane cell. All relations between the basis vectors of the two cells are expressed as vector equations, *i.e.* $\mathbf{a}'$ and $\mathbf{b}'$ are given as linear combinations of $\mathbf{a}$, $\mathbf{b}$ and $\mathbf{c}$. For the triclinic or monoclinic space groups, basis vectors $\mathbf{a}$, $\mathbf{b}$ or $\mathbf{c}$ inclined to the plane of projection are replaced by the projected vectors $\mathbf{a}_p$, $\mathbf{b}_p$, $\mathbf{c}_p$.

For primitive three-dimensional cells, the *metrical* relations between the lattice parameters of the space group and the plane group are collected in Table 2.2.14.1. The additional relations for centred cells can be derived easily from the table.

(iv) *Location of the origin* of the plane group with respect to the unit cell of the space group. The same description is used as for the location of symmetry elements (*cf.* Section 2.2.9).

Example
  'Origin at $x, 0, 0$' or 'Origin at $\frac{1}{4}, \frac{1}{4}, z$'.

2.2.14.2. *Projections of centred cells (lattices)*

For centred lattices, two different cases may occur:

(i) The projection direction is parallel to a lattice-centring vector. In this case, the projected plane cell is primitive for the centring types A, B, C, I and R. For F lattices, the multiplicity is reduced from 4 to 2 because c-centred plane cells result from projections along face diagonals of three-dimensional F cells.

Examples
(1) A body-centred lattice with centring vector $\frac{1}{2}(\mathbf{a} + \mathbf{b} + \mathbf{c})$ gives a primitive net, if projected along $[111]$, $[\bar{1}11]$, $[1\bar{1}1]$ or $[11\bar{1}]$.
(2) A C-centred lattice projects to a primitive net along the directions $[110]$ and $[1\bar{1}0]$.
(3) An R-centred lattice described with 'hexagonal axes' (triple cell) results in a primitive net, if projected along $[\bar{1}11]$, $[211]$ or $[1\bar{2}1]$ for the obverse setting. For the reverse setting, the corresponding directions are $[1\bar{1}1]$, $[\bar{2}\bar{1}1]$, $[121]$; *cf.* Chapter 1.2.

(ii) The projection direction is not parallel to a lattice-centring vector (general projection direction). In this case, the plane cell has the same multiplicity as the three-dimensional cell. Usually, however, this centred plane cell is unconventional and a transformation is required to obtain the conventional plane cell. This transformation has been carried out for the projection data in this volume.

Examples
(1) Projection along $[010]$ of a cubic I-centred cell leads to an unconventional quadratic c-centred plane cell. A simple cell transformation leads to the conventional quadratic p cell.
(2) Projection along $[010]$ of an orthorhombic I-centred cell leads to a rectangular c-centred plane cell, which is conventional.
(3) Projection along $[001]$ of an R-centred cell (both in obverse and reverse setting) results in a triple hexagonal plane cell h (the two-dimensional analogue of the H cell, *cf.* Chapter 1.2). A simple cell transformation leads to the conventional hexagonal p cell.

2.2.14.3. *Projections of symmetry elements*

A symmetry element of a space group does not project as a symmetry element unless its orientation bears a special relation to the projection direction; all translation components of a symmetry

Table 2.2.14.2. *Projections of crystallographic symmetry elements*

Symmetry element in three dimensions	Symmetry element in projection
Arbitrary orientation	
Symmetry centre $\bar{1}$ Rotoinversion axis $\bar{3} \equiv 3 \times \bar{1}$ $\Big\}$	Rotation point 2 (at projection of centre)
Parallel to projection direction	
Rotation axis 2; 3; 4; 6	Rotation point 2; 3; 4; 6
Screw axis 2_1	Rotation point 2
$3_1, 3_2$	3
$4_1, 4_2, 4_3$	4
$6_1, 6_2, 6_3, 6_4, 6_5$	6
Rotoinversion axis $\bar{4}$	Rotation point 4
$\bar{6} \equiv 3/m$	3, with overlap of atoms
$\bar{3} \equiv 3 \times \bar{1}$	6
Reflection plane m	Reflection line m
Glide plane with $\perp$ component*	Glide line g
Glide plane without $\perp$ component*	Reflection line m
Normal to projection direction	
Rotation axis 2; 4; 6	Reflection line m
3	None
Screw axis 4_2; $6_2, 6_4$	Reflection line m
2_1; $4_1, 4_3$; $6_1, 6_3, 6_5$	Glide line g
$3_1, 3_2$	None
Rotoinversion axis $\bar{4}$	Reflection line m parallel to axis
$\bar{6} \equiv 3/m$	Reflection line m perpendicular to axis (through projection of inversion point)
$\bar{3} \equiv 3 \times \bar{1}$	Rotation point 2 (at projection of centre)
Reflection plane m	None, but overlap of atoms
Glide plane with glide vector $\mathbf{t}$	Translation with translation vector $\mathbf{t}$

* The term 'with $\perp$ component' refers to the component of the glide vector normal to the projection direction.

operation along the projection direction vanish, whereas those perpendicular to the projection direction (*i.e.* parallel to the plane of projection) may be retained. This is summarized in Table 2.2.14.2 for the various crystallographic symmetry elements. From this table the following conclusions can be drawn:

(i) n-fold rotation axes and n-fold screw axes, as well as rotoinversion axes $\bar{4}$, *parallel to the projection direction* project as n-fold rotation points; a $\bar{3}$ axis projects as a sixfold, a $\bar{6}$ axis as a threefold rotation point. For the latter, a doubling of the projected electron density occurs owing to the mirror plane normal to the projection direction ($\bar{6} \equiv 3/m$).

(ii) n-fold rotation axes and n-fold screw axes *normal to the projection direction* (*i.e.* parallel to the plane of projection) do not project as symmetry elements if n is odd. If n is even, all rotation and rotoinversion axes project as mirror lines: the same applies to the screw axes $4_2, 6_2$ and 6_4 because they contain an axis 2. Screw axes $2_1, 4_1, 4_3, 6_1, 6_3$ and 6_5 project as glide lines because they contain 2_1.

(iii) Reflection planes *normal* to the projection direction do not project as symmetry elements but lead to a doubling of the projected electron density owing to overlap of atoms. Projection of a glide plane results in an additional translation; the new translation vector is equal to the glide vector of the glide plane. Thus, a reduction of the translation period in that particular direction takes place.

(iv) Reflection planes *parallel* to the projection direction project as reflection lines. Glide planes project as glide lines or as reflection lines, depending upon whether the glide vector has or has not a component parallel to the projection plane.

(v) Centres of symmetry, as well as $\bar{3}$ axes in *arbitrary* orientation, project as twofold rotation points.

Example: $C12/c1$ (15, *b* unique, cell choice 1)

The *C*-centred cell has lattice points at 0, 0, 0 and $\frac{1}{2}, \frac{1}{2}, 0$. In all projections, the centre $\bar{1}$ projects as a twofold rotation point. Projection along [001]: The plane cell is centred; $2 \parallel [010]$ projects as *m*; the glide component $(0, 0, \frac{1}{2})$ of glide plane *c* vanishes and thus *c* projects as *m*.

Result: Plane group *c2mm* (9), $\mathbf{a}' = \mathbf{a}_p, \mathbf{b}' = \mathbf{b}$.

Projection along [100]: The periodicity along *b* is halved because of the *C* centring; $2 \parallel [010]$ projects as *m*; the glide component $(0, 0, \frac{1}{2})$ of glide plane *c* is retained and thus *c* projects as *g*.

Result: Plane group *p2gm* (7), $\mathbf{a}' = \mathbf{b}/2, \mathbf{b}' = \mathbf{c}_p$.

Projection along [010]: The periodicity along *a* is halved because of the *C* centring; that along *c* is halved owing to the glide component $(0, 0, \frac{1}{2})$ of glide plane *c*; $2 \parallel [010]$ projects as 2.

Result: Plane group *p2* (2), $\mathbf{a}' = \mathbf{c}/2, \mathbf{b}' = \mathbf{a}/2$.

Further details about the geometry of projections can be found in publications by Buerger (1965) and Biedl (1966).

2.2.15. Maximal subgroups and minimal supergroups

The present section gives a brief summary, without theoretical explanations, of the sub- and supergroup data in the space-group tables. The theoretical background is provided in Section 8.3.3 and Part 13. Detailed sub- and supergroup data are given in *International Tables for Crystallography* Volume A1 (2004).

2.2.15.1. *Maximal non-isomorphic subgroups**

The maximal non-isomorphic subgroups $\mathcal{H}$ of a space group $\mathcal{G}$ are divided into two types:

I *translationengleiche* or *t* subgroups

II *klassengleiche* or *k* subgroups.

For practical reasons, type **II** is subdivided again into two blocks:

IIa the conventional cells of $\mathcal{G}$ and $\mathcal{H}$ are the same

IIb the conventional cell of $\mathcal{H}$ is larger than that of $\mathcal{G}$. †

Block **IIa** has no entries for space groups $\mathcal{G}$ with a primitive cell. For space groups $\mathcal{G}$ with a centred cell, it contains those maximal subgroups $\mathcal{H}$ that have lost some or all centring translations of $\mathcal{G}$ but none of the integral translations ('decentring' of a centred cell).

Within each block, the subgroups are listed in order of increasing index [*i*] and in order of decreasing space-group number for each value of *i*.

(i) *Blocks* **I** *and* **IIa**

In blocks **I** and **IIa**, *every* maximal subgroup $\mathcal{H}$ of a space group $\mathcal{G}$ is listed with the following information:

[*i*] HMS1 (HMS2, No.) Sequence of numbers.

---

* Space groups with different space-group numbers are non-isomorphic, except for the members of the 11 pairs of enantiomorphic space groups which are isomorphic.
† Subgroups belonging to the enantiomorphic space-group type of $\mathcal{G}$ are isomorphic to $\mathcal{G}$ and, therefore, are listed under **IIc** and not under **IIb**.

The symbols have the following meaning:

[*i*]: index of $\mathcal{H}$ in $\mathcal{G}$ (*cf.* Section 8.1.6, footnote);

HMS1: Hermann–Mauguin symbol of $\mathcal{H}$, referred to the coordinate system and setting of $\mathcal{G}$; this symbol may be unconventional;

(HMS2, No.): conventional short Hermann–Mauguin symbol of $\mathcal{H}$, given only if HMS1 is not in conventional short form, and the space-group number of $\mathcal{H}$.

Sequence of numbers: coordinate triplets of $\mathcal{G}$ retained in $\mathcal{H}$. The numbers refer to the numbering scheme of the coordinate triplets of the general position of $\mathcal{G}$ (*cf.* Section 2.2.9). The following abbreviations are used:

Block **I** (all translations retained):

Number +	Coordinate triplet given by *Number*, plus those obtained by adding all centring translations of $\mathcal{G}$.
(*Numbers*) +	The same, but applied to all *Numbers* between parentheses.

Block **IIa** (not all translations retained):

Number + (t_1, t_2, t_3)	Coordinate triplet obtained by adding the translation t_1, t_2, t_3 to the triplet given by *Number*.
(*Numbers*) + (t_1, t_2, t_3)	The same, but applied to all *Numbers* between parentheses.

In blocks **I** and **IIa**, sets of conjugate subgroups are linked by left-hand braces. For an example, see space group $R\bar{3}$ (148) below.

Examples

(1) $\mathcal{G}$: $C1m1$ (8)

I	[2] $C1$ ($P1$, 1)	$1+$
IIa	[2] $P1a1$ (Pc, 7)	$1; 2 + (1/2, 1/2, 0)$
	[2] $P1m1$ (Pm, 6)	$1; 2$

where the numbers have the following meaning:

$1+$	$x, y, z;\quad x + 1/2, y + 1/2, z$
$1; 2$	$x, y, z;\quad x, \bar{y}, z$
$1; 2 + (1/2, 1/2, 0)$	$x, y, z;\quad x + 1/2, \bar{y} + 1/2, z.$

(2) $\mathcal{G}$: $Fdd2$ (43)

I	[2] $F112$ ($C2$, 5)	$(1; 2)+$

where the numbers have the following meaning:

$$(1; 2)+ \quad x, y, z;\quad x + 1/2, y + 1/2, z;$$
$$x + 1/2, y, z + 1/2;\quad x, y + 1/2, z + 1/2;$$
$$\bar{x}, \bar{y}, z;\quad \bar{x} + 1/2, \bar{y} + 1/2, z;$$
$$\bar{x} + 1/2, \bar{y}, z + 1/2;\quad \bar{x}, \bar{y} + 1/2, z + 1/2.$$

(3) $\mathcal{G}$: $P4_2/nmc = P4_2/n2_1/m2/c$ (137)

I	[2] $P2/n2_1/m1$ ($Pmmn$, 59)	$1; 2; 5; 6; 9; 10; 13; 14.$

Operations $4_2, 2$ and *c*, occurring in the Hermann–Mauguin symbol of $\mathcal{G}$, are lacking in $\mathcal{H}$. In the unconventional 'tetragonal version' $P2/n2_1/m1$ of the symbol of $\mathcal{H}$, $2_1/m$ stands for two sets of $2_1/m$ (along the two orthogonal secondary symmetry directions), implying that $\mathcal{H}$ is orthorhombic. In the conventional 'orthorhombic version', the full symbol of $\mathcal{H}$ reads $P2_1/m2_1/m2/n$ and the short symbol *Pmmn*.

(ii) *Block* **IIb**

Whereas in blocks **I** and **IIa** *every* maximal subgroup $\mathcal{H}$ of $\mathcal{G}$ is listed, *this is no longer the case* for the entries of block **IIb**. The information given in this block is:

[*i*] HMS1 (Vectors) (HMS2, No.)

The symbols have the following meaning:

[*i*]: index of $\mathcal{H}$ in $\mathcal{G}$;

HMS1: Hermann–Mauguin symbol of $\mathcal{H}$, referred to the coordinate system and setting of $\mathcal{G}$; this symbol may be unconventional;*

(Vectors): basis vectors $\mathbf{a}'$, $\mathbf{b}'$, $\mathbf{c}'$ of $\mathcal{H}$ in terms of the basis vectors $\mathbf{a}$, $\mathbf{b}$, $\mathbf{c}$ of $\mathcal{G}$. No relations are given for unchanged axes, *e.g.* $\mathbf{a}' = \mathbf{a}$ is not stated;

(HMS2, No.): conventional short Hermann–Mauguin symbol, given only if HMS1 is not in conventional short form, and the space-group number of $\mathcal{H}$.

In addition to the general rule of increasing index [*i*] and decreasing space-group number (No.), the sequence of the **IIb** subgroups also depends on the type of cell enlargement. Subgroups with the same index and the same kind of cell enlargement are listed together in decreasing order of space-group number (see example 1 below).

In contradistinction to blocks **I** and **IIa**, for block **IIb** the coordinate triplets retained in $\mathcal{H}$ are *not* given. This means that the entry is the same for all subgroups $\mathcal{H}$ that have the same Hermann–Mauguin symbol and the same basis-vector relations to $\mathcal{G}$, but contain different sets of coordinate triplets. Thus, in block **IIb**, one entry may correspond to more than one subgroup,† as illustrated by the following examples.

Examples

(1) $\mathcal{G}$: *Pmm*2 (25)

  IIb ... [2] *Pbm*2 ($\mathbf{b}' = 2\mathbf{b}$) (*Pma*2, 28); [2] *Pcc*2 ($\mathbf{c}' = 2\mathbf{c}$) (27);

  ... [2] *Cmm*2 ($\mathbf{a}' = 2\mathbf{a}, \mathbf{b}' = 2\mathbf{b}$) (35); ...

Each of the subgroups is referred to its own distinct basis $\mathbf{a}'$, $\mathbf{b}'$, $\mathbf{c}'$, which is different in each case. Apart from the translations of the enlarged cell, the generators of the subgroups, referred to $\mathbf{a}'$, $\mathbf{b}'$, $\mathbf{c}'$, are as follows:

*Pbm*2	x,y,z;	$\bar{x},\bar{y},z$;	$x,\bar{y}+1/2,z$	or
	x,y,z;	$\bar{x},\bar{y}+1/2,z$;	$x,\bar{y},z$	
*Pcc*2	x,y,z;	$\bar{x},\bar{y},z$;	$x,\bar{y},z+1/2$	
*Cmm*2	x,y,z;	$x+1/2,y+1/2,z$;	$\bar{x},\bar{y},z$;	$x,\bar{y},z$ or
	x,y,z;	$x+1/2,y+1/2,z$;	$\bar{x},\bar{y},z$;	$x,\bar{y}+1/2,z$ or
	x,y,z;	$x+1/2,y+1/2,z$;	$\bar{x},\bar{y}+1/2,z$;	$x,\bar{y},z$ or
	x,y,z;	$x+1/2,y+1/2,z$;	$\bar{x},\bar{y}+1/2,z$;	$x,\bar{y}+1/2,z$.

There are thus 2, 1 or 4 actual subgroups that obey the same basis-vector relations. The difference between the several subgroups represented by one entry is due to the different sets of symmetry operations of $\mathcal{G}$ that are retained in $\mathcal{H}$. This can also be expressed as different conventional origins of $\mathcal{H}$ with respect to $\mathcal{G}$.

(2) $\mathcal{G}$: *P*3*m*1 (156)

  IIb ... [3] *H*3*m*1 ($\mathbf{a}' = 3\mathbf{a}, \mathbf{b}' = 3\mathbf{b}$) (*P*31*m*, 157)

The nine subgroups of type *P*31*m* may be described in two ways:

  (i) By partial 'decentring' of ninetuple cells ($\mathbf{a}' = 3\mathbf{a}$, $\mathbf{b}' = 3\mathbf{b}$, $\mathbf{c}' = \mathbf{c}$) with the same orientations as the cell of the group $\mathcal{G}(\mathbf{a}, \mathbf{b}, \mathbf{c})$ in such a way that the centring points $0, 0, 0$; $2/3, 1/3, 0$; $1/3, 2/3, 0$ (referred to $\mathbf{a}', \mathbf{b}', \mathbf{c}'$) are retained. The *conventional* space-group symbol *P*31*m* of these nine subgroups is referred to the same basis vectors $\mathbf{a}'' = \mathbf{a} - \mathbf{b}$, $\mathbf{b}'' = \mathbf{a} + 2\mathbf{b}$, $\mathbf{c}'' = \mathbf{c}$, but to different origins; *cf.* Section 2.2.15.5. This kind of description is used in the space-group tables of this volume.

  (ii) Alternatively, one can describe the group $\mathcal{G}$ with an unconventional *H*-centred cell ($\mathbf{a}' = \mathbf{a} - \mathbf{b}$, $\mathbf{b}' = \mathbf{a} + 2\mathbf{b}$, $\mathbf{c}' = \mathbf{c}$) referred to which the space-group symbol is *H*31*m*. 'Decentring' of this cell results in the conventional space-group symbol *P*31*m* for the subgroups, referred to the basis vectors $\mathbf{a}'$, $\mathbf{b}'$, $\mathbf{c}'$. This description is used in Section 4.3.5.

(iii) *Subdivision of k subgroups into blocks* **IIa** *and* **IIb**

The subdivision of *k* subgroups into blocks **IIa** and **IIb** has no group-theoretical background and depends on the coordinate system chosen. The *conventional* coordinate system of the space group $\mathcal{G}$ (*cf.* Section 2.1.3) is taken as the basis for the subdivision. This results in a uniquely defined subdivision, except for the seven rhombohedral space groups for which in the space-group tables both 'rhombohedral axes' (primitive cell) and 'hexagonal axes' (triple cell) are given (*cf.* Section 2.2.2). Thus, some *k* subgroups of a rhombohedral space group are found under **IIa** (*klassengleich*, centring translations lost) in the *hexagonal* description, and under **IIb** (*klassengleich*, conventional cell enlarged) in the *rhombohedral* description.

Example: $\mathcal{G}$: $R\bar{3}$ (148) $\mathcal{H}$: $P\bar{3}$ (147)
  Hexagonal axes

I	[2] *R*3 (146)	$(1;2;3)+$	
	[3] $R\bar{1}$ ($P\bar{1}$, 2)	$(1;4)+$	
IIa	[3] $P\bar{3}$ (147)	$1;2;3;4;5;6$	
	[3] $P\bar{3}$ (147)	$1;2;3;(4;5;6) + (\frac{1}{3}, \frac{2}{3}, \frac{2}{3})$	
	[3] $P\bar{3}$ (147)	$1;2;3;(4;5;6) + (\frac{2}{3}, \frac{1}{3}, \frac{1}{3})$	
IIb	none		

Rhombohedral axes

I	[2] *R*3 (146)	$1;2;3$
	[3] $R\bar{1}$ ($P\bar{1}$, 2)	$1;4$
IIa	none	
IIb	[3] $P\bar{3}$ ($\mathbf{a}' = \mathbf{a} - \mathbf{b}, \mathbf{b}' = \mathbf{b} - \mathbf{c}, \mathbf{c}' = \mathbf{a} + \mathbf{b} + \mathbf{c}$) (147).

Apart from the change from **IIa** to **IIb**, the above example demonstrates again the restricted character of the **IIb** listing, discussed above. The three conjugate subgroups $P\bar{3}$ of index [3] are listed under **IIb** by one entry only, because for all three subgroups the basis-vector relations between $\mathcal{G}$ and $\mathcal{H}$ are the same. Note the brace for the **IIa** subgroups, which unites *conjugate subgroups* into classes.

2.2.15.2. *Maximal isomorphic subgroups of lowest index (cf. Part 13)*

Another set of *klassengleiche* subgroups are the *isomorphic subgroups* listed under **IIc**, *i.e.* the subgroups $\mathcal{H}$ which are of the

---

* Unconventional Hermann–Mauguin symbols may include unconventional cells like *c* centring in quadratic plane groups, *F* centring in monoclinic, or *C* and *F* centring in tetragonal space groups. Furthermore, the triple hexagonal cells *h* and *H* are used for certain sub- and supergroups of the hexagonal plane groups and of the trigonal and hexagonal *P* space groups, respectively. The cells *h* and *H* are defined in Chapter 1.2. Examples are subgroups of plane groups *p*3 (13) and *p*6*mm* (17) and of space groups *P*3 (143) and *P*6/*mcc* (192).

† Without this restriction, the amount of data would be excessive. For instance, space group *Pmmm* (47) has 63 maximal subgroups of index [2], of which seven are *t* subgroups and listed explicitly under **I**. The 16 entries under **IIb** refer to 50 actual subgroups and the one entry under **IIc** stands for the remaining 6 subgroups.

same or of the enantiomorphic space-group type as $\mathcal{G}$. The kind of listing is the same as for block **IIb**. Again, one entry may correspond to more than one isomorphic subgroup.

As the number of maximal isomorphic subgroups of a space group is always infinite, the data in block **IIc** are restricted to the subgroups of lowest index. Different kinds of cell enlargements are presented. For monoclinic, tetragonal, trigonal and hexagonal space groups, cell enlargements both parallel and perpendicular to the main rotation axis are listed; for orthorhombic space groups, this is the case for all three directions, a, b and c. Two isomorphic subgroups $\mathcal{H}_1$ and $\mathcal{H}_2$ of equal index but with cell enlargements in different directions may, nevertheless, play an analogous role with respect to $\mathcal{G}$. In terms of group theory, $\mathcal{H}_1$ and $\mathcal{H}_2$ then are conjugate subgroups in the affine normalizer of $\mathcal{G}$, *i.e.* they are mapped onto each other by automorphisms of $\mathcal{G}$.* Such subgroups are collected into one entry, with the different vector relationships separated by 'or' and placed within one pair of parentheses; *cf.* example (4).

Examples
(1) $\mathcal{G}$: $P\bar{3}1c$ (163)

    IIc [3] $P\bar{3}1c$ ($\mathbf{c}' = 3\mathbf{c}$) (163); [4] $P\bar{3}1c$ ($\mathbf{a}' = 2\mathbf{a}, \mathbf{b}' = 2\mathbf{b}$) (163).

The first subgroup of index [3] entails an enlargement of the c axis, the second one of index [4] an enlargement of the mesh size in the a,b plane.
(2) $\mathcal{G}$: $P23$ (195)

    IIc [27] $P23$ ($\mathbf{a}' = 3\mathbf{a}, \mathbf{b}' = 3\mathbf{b}, \mathbf{c}' = 3\mathbf{c}$) (195).

It seems surprising that [27] is the lowest index listed, even though another isomorphic subgroup of index [8] exists. The latter subgroup, however, is not maximal, as chains of maximal non-isomorphic subgroups can be constructed as follows:

$$P23 \to [4]\ I23\ (\mathbf{a}' = 2\mathbf{a}, \mathbf{b}' = 2\mathbf{b}, \mathbf{c}' = 2\mathbf{c}) \to [2]\ P23\ (\mathbf{a}', \mathbf{b}', \mathbf{c}')$$

or

$$P23 \to [2]\ F23\ (\mathbf{a}' = 2\mathbf{a}, \mathbf{b}' = 2\mathbf{b}, \mathbf{c}' = 2\mathbf{c}) \to [4]\ P23\ (\mathbf{a}', \mathbf{b}', \mathbf{c}').$$

(3) $\mathcal{G}$: $P3_112$ (151)

    IIc [2] $P3_212$ ($\mathbf{c}' = 2\mathbf{c}$) (153); [4] $P3_112$ ($\mathbf{a}' = 2\mathbf{a}, \mathbf{b}' = 2\mathbf{b}$) (151);

        [7] $P3_112$ ($\mathbf{c}' = 7\mathbf{c}$) (151).

Note that the isomorphic subgroup of index [4] with $\mathbf{c}' = 4\mathbf{c}$ is not listed, because it is not maximal. This is apparent from the chain

$$P3_112 \to [2]\ P3_212\ (\mathbf{c}' = 2\mathbf{c}) \to [2]\ P3_112\ (\mathbf{c}'' = 2\mathbf{c}' = 4\mathbf{c}).$$

(4) $\mathcal{G}_1$: $Pnnm$ (58)

    IIc [3] $Pnnm$ ($\mathbf{a}' = 3\mathbf{a}$ or $\mathbf{b}' = 3\mathbf{b}$) (58); [3] $Pnnm$ ($\mathbf{c}' = 3\mathbf{c}$) (58);

but $\mathcal{G}_2$: $Pnna$ (52)

    IIc [3] $Pnna$ ($\mathbf{a}' = 3\mathbf{a}$) (52); [3] $Pnna$ ($\mathbf{b}' = 3\mathbf{b}$) (52);

        [3] $Pnna$ ($\mathbf{c}' = 3\mathbf{c}$) (52).

For $\mathcal{G}_1 = Pnnm$, the x and y directions are analogous, *i.e.* they may be interchanged by automorphisms of $\mathcal{G}_1$. Such an automorphism does not exist for $\mathcal{G}_2 = Pnna$ because this space group contains glide reflections a but not b.

2.2.15.3. *Minimal non-isomorphic supergroups*

If $\mathcal{G}$ is a maximal subgroup of a group $\mathcal{S}$, then $\mathcal{S}$ is called a minimal supergroup of $\mathcal{G}$. Minimal non-isomorphic supergroups are

again subdivided into two types, the *translationengleiche* or t supergroups **I** and the *klassengleiche* or k supergroups **II**. For the minimal t supergroups **I** of $\mathcal{G}$, the listing contains the index [i] of $\mathcal{G}$ in $\mathcal{S}$, the *conventional* Hermann–Mauguin symbol of $\mathcal{S}$ and its space-group number in parentheses.

There are two types of minimal k supergroups **II**: supergroups with additional centring translations (which would correspond to the **IIa** type) and supergroups with smaller conventional unit cells than that of $\mathcal{G}$ (type **IIb**). Although the subdivision between **IIa** and **IIb** supergroups is not indicated in the tables, the list of minimal supergroups with additional centring translations (**IIa**) always precedes the list of **IIb** supergroups. The information given is similar to that for the non-isomorphic subgroups **IIb**, *i.e.*, where applicable, the relations between the basis vectors of group and supergroup are given, in addition to the Hermann–Mauguin symbols of $\mathcal{S}$ and its space-group number. The supergroups are listed in order of increasing index and increasing space-group number.

The block of supergroups contains only the *types* of the non-isomorphic minimal supergroups $\mathcal{S}$ of $\mathcal{G}$, *i.e.* each entry may correspond to more than one supergroup $\mathcal{S}$. In fact, the list of minimal supergroups $\mathcal{S}$ of $\mathcal{G}$ should be considered as a backwards reference to those space groups $\mathcal{S}$ for which $\mathcal{G}$ appears as a maximal subgroup. Thus, the relation between $\mathcal{S}$ and $\mathcal{G}$ can be found in the subgroup entries of $\mathcal{S}$.

Example: $\mathcal{G}$: $Pna2_1$ (33)
    Minimal non-isomorphic supergroups

    I   [2] $Pnna$ (52); [2] $Pccn$ (56); [2] $Pbcn$ (60); [2] $Pnma$ (62).

    II ...[2] $Pnm2_1$ ($\mathbf{a}' = \frac{1}{2}\mathbf{a}$) ($Pmn2_1, 31$); ....

Block **I** lists, among others, the entry [2] $Pnma$ (62). Looking up the *subgroup* data of $Pnma$ (62), one finds in block **I** the entry [2] $Pn2_1a$ ($Pna2_1$). This shows that the setting of $Pnma$ does not correspond to that of $Pna2_1$ but rather to that of $Pn2_1a$. To obtain the supergroup $\mathcal{S}$ referred to the basis of $Pna2_1$, the basis vectors $\mathbf{b}$ and $\mathbf{c}$ must be interchanged. This changes $Pnma$ to $Pnam$, which is the correct symbol of the supergroup of $Pna2_1$.

Note on R supergroups of trigonal P space groups: The trigonal P space groups Nos. 143–145, 147, 150, 152, 154, 156, 158, 164 and 165 each have two rhombohedral supergroups of type **II**. They are distinguished by different additional centring translations which correspond to the 'obverse' and 'reverse' settings of a triple hexagonal R cell; *cf.* Chapter 1.2. In the supergroup tables of Part 7, these cases are described as [3] $R3$ (obverse) (146); [3] $R3$ (reverse) (146) *etc.*

2.2.15.4. *Minimal isomorphic supergroups of lowest index*

No data are listed for isomorphic *supergroups* **IIc** because they can be derived directly from the corresponding data of *subgroups* **IIc** (*cf.* Part 13).

2.2.15.5. *Note on basis vectors*

In the *subgroup* data, $\mathbf{a}'$, $\mathbf{b}'$, $\mathbf{c}'$ are the basis vectors of the subgroup $\mathcal{H}$ of the space group $\mathcal{G}$. The latter has the basis vectors $\mathbf{a}$, $\mathbf{b}$, $\mathbf{c}$. In the *supergroup* data, $\mathbf{a}'$, $\mathbf{b}'$, $\mathbf{c}'$ are the basis vectors of the supergroup $\mathcal{S}$ and $\mathbf{a}$, $\mathbf{b}$, $\mathbf{c}$ are again the basis vectors of $\mathcal{G}$. Thus, $\mathbf{a}$, $\mathbf{b}$, $\mathbf{c}$ and $\mathbf{a}'$, $\mathbf{b}'$, $\mathbf{c}'$ exchange their roles if one considers the same group–subgroup relation in the subgroup and the supergroup tables.

Examples
(1) $\mathcal{G}$: $Pba2$ (32)

    Listed under *subgroups* **IIb**, one finds, among other entries, [2] $Pna2_1$ ($\mathbf{c}' = 2\mathbf{c}$) (33); thus, $\mathbf{c}(Pna2_1) = 2\mathbf{c}(Pba2)$.

---

* For normalizers of space groups, see Section 8.3.6 and Part 15, where also references to automorphisms are given.

Under *supergroups* **II** of *Pna2₁* (33), the corresponding entry reads [2] *Pba2* ($\mathbf{c}' = \frac{1}{2}\mathbf{c}$) (32); thus $\mathbf{c}(Pba2) = \frac{1}{2}\mathbf{c}(Pna2_1)$.

(2) Tetragonal *k* space groups with *P* cells. For index [2], the relations between the *conventional* basis vectors of the group and the subgroup read (*cf.* Fig. 5.1.3.5)

$$\mathbf{a}' = \mathbf{a} + \mathbf{b}, \qquad \mathbf{b}' = -\mathbf{a} + \mathbf{b} \qquad (\mathbf{a}', \mathbf{b}' \text{ for the subgroup}).$$

Thus, the basis vectors of the supergroup are

$$\mathbf{a}' = \tfrac{1}{2}(\mathbf{a} - \mathbf{b}), \quad \mathbf{b}' = \tfrac{1}{2}(\mathbf{a} + \mathbf{b}) \qquad (\mathbf{a}', \mathbf{b}' \text{ for the supergroup}).$$

An alternative description is

$$\mathbf{a}' = \mathbf{a} - \mathbf{b}, \qquad \mathbf{b}' = \mathbf{a} + \mathbf{b} \qquad (\mathbf{a}', \mathbf{b}' \text{ for the subgroup})$$
$$\mathbf{a}' = \tfrac{1}{2}(\mathbf{a} + \mathbf{b}), \quad \mathbf{b}' = \tfrac{1}{2}(-\mathbf{a} + \mathbf{b}) \quad (\mathbf{a}', \mathbf{b}' \text{ for the supergroup}).$$

(3) Hexagonal *k* space groups. For index [3], the relations between the *conventional* basis vectors of the sub- and supergroup read (*cf.* Fig 5.1.3.8)

$$\mathbf{a}' = \mathbf{a} - \mathbf{b}, \qquad \mathbf{b}' = \mathbf{a} + 2\mathbf{b} \qquad (\mathbf{a}', \mathbf{b}' \text{ for the subgroup}).$$

Thus, the basis vectors of the supergroup are

$$\mathbf{a}' = \tfrac{1}{3}(2\mathbf{a} + \mathbf{b}), \quad \mathbf{b}' = \tfrac{1}{3}(-\mathbf{a} + \mathbf{b}) \quad (\mathbf{a}', \mathbf{b}' \text{ for the supergroup}).$$

An alternative description is

$$\mathbf{a}' = 2\mathbf{a} + \mathbf{b}, \qquad \mathbf{b}' = -\mathbf{a} + \mathbf{b} \qquad (\mathbf{a}', \mathbf{b}' \text{ for the subgroup})$$
$$\mathbf{a}' = \tfrac{1}{3}(\mathbf{a} - \mathbf{b}), \quad \mathbf{b}' = \tfrac{1}{3}(\mathbf{a} + 2\mathbf{b}) \quad (\mathbf{a}', \mathbf{b}' \text{ for the supergroup}).$$

2.2.16. Monoclinic space groups

In this volume, space groups are described by one (or at most two) conventional coordinate systems (*cf.* Sections 2.1.3 and 2.2.2). Eight monoclinic space groups, however, are treated more extensively. In order to provide descriptions for frequently encountered cases, they are given in six versions.

The description of a monoclinic crystal structure in this volume, including its Hermann–Mauguin space-group symbol, depends upon two choices:

  (i) the unit cell chosen, here called 'cell choice';

  (ii) the labelling of the edges of this cell, especially of the monoclinic symmetry direction ('unique axis'), here called 'setting'.

2.2.16.1. Cell choices

One edge of the cell, *i.e.* one crystal axis, is always chosen along the monoclinic symmetry direction. The other two edges are located in the plane perpendicular to this direction and coincide with translation vectors in this 'monoclinic plane'. It is sensible and common practice (see below) to choose these two basis vectors from the *shortest three* translation vectors in that plane. They are shown in Fig. 2.2.16.1 and labelled **e**, **f** and **g**, in order of increasing length.* The two shorter vectors span the 'reduced mesh', here **e** and **f**; for this mesh, the monoclinic angle is $\leq 120°$, whereas for the other two primitive meshes larger angles are possible.

Other choices of the basis vectors in the monoclinic plane are possible, provided they span a primitive mesh. It turns out, however, that the space-group symbol for any of these (non-reduced) meshes already occurs among the symbols for the three meshes formed by **e**, **f**, **g** in Fig. 2.2.16.1; hence only these cases need be considered. They are designated in this volume as 'cell choice 1, 2 or 3' and are depicted in Fig. 2.2.6.4. The transformation matrices for the three cell choices are listed in Table 5.1.3.1.

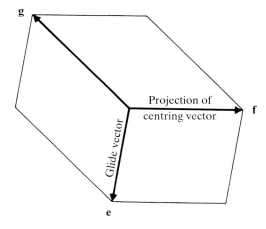

Fig. 2.2.16.1. The three primitive two-dimensional cells which are spanned by the shortest three translation vectors **e**, **f**, **g** in the monoclinic plane. For the present discussion, the glide vector is considered to be along **e** and the projection of the centring vector along **f**.

2.2.16.2. Settings

The term *setting* of a cell or of a space group refers to the assignment of labels (*a*, *b*, *c*) and directions to the edges of a given unit cell, resulting in a set of basis vectors **a**, **b**, **c**. (For orthorhombic space groups, the six settings are described and illustrated in Section 2.2.6.4.)

The symbol for each setting is a shorthand notation for the transformation of a given starting set **abc** into the setting considered. It is called here 'setting symbol'. For instance, the setting symbol **bca** stands for

$$\mathbf{a}' = \mathbf{b}, \quad \mathbf{b}' = \mathbf{c}, \quad \mathbf{c}' = \mathbf{a}$$

or

$$(\mathbf{a}'\mathbf{b}'\mathbf{c}') = (\mathbf{abc}) \begin{pmatrix} 0 & 0 & 1 \\ 1 & 0 & 0 \\ 0 & 1 & 0 \end{pmatrix} = (\mathbf{bca}),$$

where **a**′, **b**′, **c**′ is the new set of basis vectors. (Note that the setting symbol **bca** does *not* mean that the old vector **a** changes its label to **b**, the old vector **b** changes to **c**, and the old **c** changes to **a**.) Transformation of one setting into another preserves the shape of the cell and its orientation relative to the lattice. The matrices of these transformations have *one* entry +1 or −1 in each row and column; all other entries are 0.

In monoclinic space groups, one axis, the monoclinic symmetry direction, is unique. Its label must be chosen first and, depending upon this choice, one speaks of 'unique axis *b*', 'unique axis *c*' or 'unique axis *a*'.† Conventionally, the positive directions of the two further ('oblique') axes are oriented so as to make the monoclinic angle non-acute, *i.e.* $\geq 90°$, and the coordinate system right-handed. For the three cell choices, settings obeying this condition and having the same label and direction of the unique axis are considered as one setting; this is illustrated in Fig. 2.2.6.4.

Note: These three cases of labelling the monoclinic axis are often called somewhat loosely *b*-axis, *c*-axis and *a*-axis 'settings'. It must be realized, however, that the choice of the 'unique axis' alone does *not* define a *single* setting but only a *pair*, as for each cell the labels of the two oblique axes can be interchanged.

---

* These three vectors obey the 'closed-triangle' condition $\mathbf{e} + \mathbf{f} + \mathbf{g} = \mathbf{0}$; they can be considered as two-dimensional homogeneous axes.

† In *IT* (1952), the terms '1st setting' and '2nd setting' were used for 'unique axis *c*' and 'unique axis *b*'. In the present volume, these terms have been dropped in favour of the latter names, which are unambiguous.

Table 2.2.16.1 lists the setting symbols for the six monoclinic settings in three equivalent forms, starting with the symbols **a** **b** **c** (first line), **a** **b** **c** (second line) and **a** **b** **c** (third line); the unique axis is underlined. These symbols are also found in the headline of the synoptic Table 4.3.2.1, which lists the space-group symbols for all monoclinic settings and cell choices. Again, the corresponding transformation matrices are listed in Table 5.1.3.1.

In the space-group tables, only the settings with b and c unique are treated and for these only the left-hand members of the double entries in Table 2.2.16.1. This implies, for instance, that the c-axis setting is obtained from the b-axis setting by cyclic permutation of the labels, $i.e.$ by the transformation

$$(\mathbf{a'b'\underline{c}'}) = (\mathbf{a\underline{b}c})\begin{pmatrix} 0 & 1 & 0 \\ 0 & 0 & 1 \\ 1 & 0 & 0 \end{pmatrix} = (\mathbf{ca\underline{b}}).$$

In the present discussion, also the setting with a unique is included, as this setting occurs in the subgroup entries of Part 7 and in Table 4.3.2.1. The a-axis setting $\mathbf{\underline{a}'b'c'} = \mathbf{\underline{c}ab}$ is obtained from the c-axis setting also by cyclic permutation of the labels and from the b-axis setting by the reverse cyclic permutation: $\mathbf{\underline{a}'b'c'} = \mathbf{\underline{b}ca}$.

By the conventions described above, the setting of each of the cell choices 1, 2 and 3 is determined once the label and the direction of the unique-axis vector have been selected. Six of the nine resulting possibilities are illustrated in Fig. 2.2.6.4.

2.2.16.3. Cell choices and settings in the present tables

There are five monoclinic space groups for which the Hermann–Mauguin symbols are independent of the cell choice, viz those space groups that do *not* contain centred lattices or glide planes:

$P2$ (No. 3), $P2_1$ (4), Pm (6), $P2/m$ (10), $P2_1/m$ (11).

In these cases, description of the space group by one cell choice is sufficient.

For the eight monoclinic space groups *with centred lattices or glide planes*, the Hermann–Mauguin symbol depends on the choice of the oblique axes with respect to the glide vector and/or the centring vector. These eight space groups are:

$C2$ (5), Pc (7), Cm (8), Cc (9), $C2/m$ (12), $P2/c$ (13),

$P2_1/c$ (14), $C2/c$ (15).

Here, the glide vector or the projection of the centring vector onto the monoclinic plane are always directed along *one* of the vectors **e**, **f** or **g** in Fig. 2.2.16.1, $i.e.$ are parallel to the shortest, the second-shortest or the third-shortest translation vector in the monoclinic plane (note that a glide vector and the projection of a centring vector cannot be parallel). This results in three possible orientations of the glide vector or the centring vector with respect to these crystal axes, and thus in three different full Hermann–Mauguin symbols ($cf.$ Section 2.2.4) for each setting of a space group.

Table 2.2.16.2 lists the symbols for centring types and glide planes for the cell choices 1, 2, 3. The order of the three cell choices is defined as follows: The symbols occurring in the familiar 'standard short monoclinic space-group symbols' (see Section 2.2.3) define cell choice 1; for 'unique axis b', this applies to the centring type C and the glide plane c, as in Cm (8) and $P2_1/c$ (14). Cell choices 2 and 3 follow from the anticlockwise order 1–2–3 in Fig. 2.2.6.4 and their space-group symbols can be obtained from Table 2.2.16.2. The c-axis and the a-axis settings then are derived from the b-axis setting by cyclic permutations of the axial labels, as described in Section 2.2.16.2.

In the two space groups Cc (9) and $C2/c$ (15), glide planes occur in pairs, $i.e.$ each vector **e**, **f**, **g** is associated either with a glide vector or with the centring vector of the cell. For Pc (7), $P2/c$ (13) and

Table 2.2.16.1. *Monoclinic setting symbols (unique axis is underlined)*

Unique axis b		Unique axis c		Unique axis a			
a$\underline{b}$c	c$\overline{\underline{b}}$a	ca$\underline{b}$	ac$\overline{\underline{b}}$	$\underline{b}$ca	$\underline{b}$ac	Starting set a$\underline{b}$c	
$\underline{b}$ca	a$\overline{\underline{c}}$b	a$\underline{b}$c	ba$\overline{\underline{c}}$	ca$\underline{b}$	$\overline{\underline{c}}$ba	Starting set a$\underline{b}$c	
c$\underline{a}$b	b$\overline{\underline{a}}$c	bc$\underline{a}$	cb$\overline{\underline{a}}$	$\underline{a}$bc	$\overline{\underline{a}}$cb	Starting set $\underline{a}$bc	

$Note$: An interchange of two axes involves a change of the handedness of the coordinate system. In order to keep the system right-handed, one sign reversal is necessary.

Table 2.2.16.2. *Symbols for centring types and glide planes of monoclinic space groups*

Setting		Cell choice		
		1	2	3
Unique axis b	Centring type	C	A	I
	Glide planes	c, n	n, a	a, c
Unique axis c	Centring type	A	B	I
	Glide planes	a, n	n, b	b, a
Unique axis a	Centring type	B	C	I
	Glide planes	b, n	n, c	c, b

$P2_1/c$ (14), which contain only one type of glide plane, the left-hand member of each pair of glide planes in Table 2.2.16.2 applies.

In the space-group tables of this volume, the following treatments of monoclinic space groups are given:

(1) *Two complete descriptions* for each of the five monoclinic space groups with primitive lattices and without glide planes, one for 'unique axis b' and one for 'unique axis c', similar to the treatment in IT (1952).

(2) A total of *six descriptions* for each of the eight space groups with centred lattices or glide planes, as follows:

(a) *One complete* description for 'unique axis b' and 'cell choice' 1. This is considered the standard description of the space group, and its *short* Hermann–Mauguin symbol is used as the *standard* symbol of the space group.

This standard short symbol corresponds to the one symbol of IT (1935) and to that of the b-axis setting in IT (1952), $e.g.$ $P2_1/c$ or $C2/c$. It serves only to identify the space-group type but carries no information about the setting or cell choice of a particular description. The *standard short symbol* is given in the headline of every description of a monoclinic space group; $cf.$ Section 2.2.3.

(b) *Three condensed* (synoptic) descriptions for 'unique axis b' and the three 'cell choices' 1, 2, 3. Cell choice 1 is repeated to facilitate comparison with the other cell choices. Diagrams are provided to illustrate the three cell choices: $cf.$ Section 2.2.6.

(c) *One complete* description for 'unique axis c' and 'cell choice' 1.

(d) *Three condensed* (synoptic) descriptions for 'unique axis c' and the three 'cell choices' 1, 2, 3. Again cell choice 1 is repeated and appropriate diagrams are provided.

All settings and cell choices are identified by the appropriate *full* Hermann–Mauguin symbols ($cf.$ Section 2.2.4), $e.g.$ $C12/c1$ or $I112/b$. For the two space groups Cc (9) and $C2/c$ (15) with pairs of different glide planes, the 'priority rule' ($cf.$ Section 4.1.1) for glide planes (*e* before *a* before *b* before *c* before *n*) is *not* followed. Instead, in order to bring out the relations between the various settings and cell choices, the glide-plane symbol always refers to that glide plane which intersects the conventional origin.

Example: No. 15, standard short symbol *C*2/*c*

The full symbols for the three cell choices (rows) and the three unique axes (columns) read

$$C12/c1 \quad A12/n1 \quad I12/a1$$
$$A112/a \quad B112/n \quad I112/b$$
$$B2/b11 \quad C2/n11 \quad I2/c11.$$

Application of the priority rule would have resulted in the following symbols

$$C12/c1 \quad A12/a1 \quad I12/a1$$
$$A112/a \quad B112/b \quad I112/a$$
$$B2/b11 \quad C2/c11 \quad I2/b11.$$

Here, the transformation properties are obscured.

2.2.16.4. *Comparison with earlier editions of International Tables*

In *IT* (1935), each monoclinic space group was presented in one description only, with *b* as the unique axis. Hence, only one short Hermann–Mauguin symbol was needed.

In *IT* (1952), the *c*-axis setting (first setting) was newly introduced, in addition to the *b*-axis setting (second setting). This extension was based on a decision of the Stockholm General Assembly of the International Union of Crystallography in 1951 [*cf. Acta Cryst.* (1951), **4**, 569 and *Preface* to *IT* (1952)]. According to this decision, the *b*-axis setting should continue to be accepted as standard for morphological and structural studies. The two settings led to the introduction of *full* Hermann–Mauguin symbols for *all 13* monoclinic space groups (*e.g.* $P12_1/c1$ and $P112_1/b$) and of two different *standard short* symbols (*e.g.* $P2_1/c$ and $P2_1/b$) for the *eight* space groups with centred lattices or glide planes [*cf.* p. 545 of *IT* (1952)]. In the present volume, only one of these standard short symbols is retained (see above and Section 2.2.3).

The *c*-axis setting (primed labels) was obtained from the *b*-axis setting (unprimed labels) by the following transformation

$$(\mathbf{a'b'\underline{c}'}) = (\mathbf{a\underline{b}c}) \begin{pmatrix} 1 & 0 & 0 \\ 0 & 0 & \bar{1} \\ 0 & 1 & 0 \end{pmatrix} = (\mathbf{ac\underline{b}}).$$

This corresponds to an interchange of two labels and not to the more logical cyclic permutation, as used in the present volume. The reason for this particular transformation was to obtain short space-group symbols that indicate the setting unambiguously; thus the lattice letters were chosen as *C* (*b*-axis setting) and *B* (*c*-axis setting). The use of *A* in either case would not have distinguished between the two settings [*cf.* pp. 7, 55 and 543 of *IT* (1952); see also Table 2.2.16.2].

As a consequence of the different transformations between *b*- and *c*-axis settings in *IT* (1952) and in this volume, some space-group symbols have changed. This is apparent from a comparison of pairs such as $P12_1/c1$ & $P112_1/b$ and $C12/c1$ & $B112/b$ in *IT* (1952) with the corresponding pairs in this volume, $P12_1/c1$ & $P112_1/a$ and $C12/c1$ & $A112/a$. The symbols with *B*-centred cells appear now for cell choice 2, as can be seen from Table 2.2.16.2.

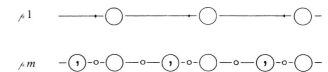

Fig. 2.2.17.1. The two line groups (one-dimensional space groups). Small circles are reflection points; large circles represent the general position; in line group $\not{p}1$, the vertical bars are the origins of the unit cells.

2.2.16.5. *Selection of monoclinic cell*

In practice, the selection of the (right-handed) unit cell of a monoclinic crystal can be approached in three ways, whereby the axes refer to the *b*-unique setting; for *c* unique similar considerations apply:

(i) Irrespective of their lengths, the basis vectors are chosen such that, in Fig. 2.2.16.1, one obtains $\mathbf{c} = \mathbf{e}$, $\mathbf{a} = \mathbf{f}$ and $\mathbf{b}$ normal to $\mathbf{a}$ and $\mathbf{c}$ pointing upwards. This corresponds to a selection of cell choice 1. It ensures that the crystal structure can always be referred directly to the description and the space-group symbol in *IT* (1935) and *IT* (1952). However, this is at the expense of possibly using a non-reduced and, in many cases, even a very awkward cell.

(ii) Selection of the reduced mesh, *i.e.* the shortest two translation vectors in the monoclinic plane are taken as axes and labelled **a** and **c**, with either $a < c$ or $c < a$. This results with equal probability in one of the three cell choices described in the present volume.

(iii) Selection of the cell on special grounds, *e.g.* to compare the structure under consideration with another related crystal structure. This may result again in a non-reduced cell and it may even necessitate use of the *a*-axis setting. In all these cases, the coordinate system chosen should be carefully explained in the description of the structure.

2.2.17. Crystallographic groups in one dimension

In one dimension, only one crystal family, one crystal system and one Bravais lattice exist. No name or common symbol is required for any of them. All one-dimensional lattices are primitive, which is symbolized by the script letter $\not{p}$; *cf.* Chapter 1.2.

There occur two types of one-dimensional point groups, 1 and $m \equiv \bar{1}$. The latter contains reflections through a point (reflection point or mirror point). This operation can also be described as inversion through a point, thus $m \equiv \bar{1}$ for one dimension; *cf.* Chapters 1.3 and 1.4.

Two types of line groups (one-dimensional space groups) exist, with Hermann–Mauguin symbols $\not{p}1$ and $\not{p}m \equiv \not{p}\bar{1}$, which are illustrated in Fig. 2.2.17.1. Line group $\not{p}1$, which consists of one-dimensional translations only, has merely one (general) position with coordinate *x*. Line group $\not{p}m$ consists of one-dimensional translations and reflections through points. It has one general and two special positions. The coordinates of the general position are *x* and $\bar{x}$; the coordinate of one special position is 0, that of the other $\frac{1}{2}$. The site symmetries of both special positions are $m \equiv \bar{1}$. For $\not{p}1$, the origin is arbitrary, for $\not{p}m$ it is at a reflection point.

The one-dimensional *point groups* are of interest as 'edge symmetries' of two-dimensional 'edge forms'; they are listed in Table 10.1.2.1. The one-dimensional *space groups* occur as projection and section symmetries of crystal structures.

References

2.1

Hahn, Th. & Wondratschek, H. (1994). *Symmetry of crystals.* Sofia: Heron Press.

Internationale Tabellen zur Bestimmung von Kristallstrukturen (1935). 1. Band, edited by C. Hermann. Berlin: Borntraeger. [Revised edition: Ann Arbor: Edwards (1944). Abbreviated as *IT* (1935).]

International Tables for Crystallography (2004). Vol. C, 3rd ed., edited by A. J. C. Wilson & E. Prince. Dordrecht: Kluwer Academic Publishers.

International Tables for X-ray Crystallography (1952). Vol. I, edited by N. F. M. Henry & K. Lonsdale. Birmingham: Kynoch Press. [Revised editions: 1965, 1969 and 1977. Abbreviated as *IT* (1952).]

Wolff, P. M. de, Belov, N. V., Bertaut, E. F., Buerger, M. J., Donnay, J. D. H., Fischer, W., Hahn, Th., Koptsik, V. A., Mackay, A. L., Wondratschek, H., Wilson, A. J. C. & Abrahams, S. C. (1985). *Nomenclature for crystal families, Bravais-lattice types and arithmetic classes. Report of the International Union of Crystallography Ad-hoc Committee on the Nomenclature of Symmetry.* Acta Cryst. A**41**, 278–280.

2.2

Astbury, W. T. & Yardley, K. (1924). *Tabulated data for the examination of the 230 space groups by homogeneous X-rays.* Philos. Trans. R. Soc. London Ser. A, **224**, 221–257.

Belov, N. V., Zagal'skaja, Ju. G., Litvinskaja, G. P. & Egorov-Tismenko, Ju. K. (1980). *Atlas of the space groups of the cubic system.* Moscow: Nauka. (In Russian.)

Biedl, A. W. (1966). *The projection of a crystal structure.* Z. Kristallogr. **123**, 21–26.

Buerger, M. J. (1949). *Fourier summations for symmetrical crystals.* Am. Mineral. **34**, 771–788.

Buerger, M. J. (1956). *Elementary crystallography.* New York: Wiley.

Buerger, M. J. (1959). *Vector space.* New York: Wiley.

Buerger, M. J. (1960). *Crystal-structure analysis*, Ch. 17. New York: Wiley.

Buerger, M. J. (1965). *The geometry of projections.* Tschermaks Mineral. Petrogr. Mitt. **10**, 595–607.

Engel, P., Matsumoto, T., Steinmann, G. & Wondratschek, H. (1984). *The non-characteristic orbits of the space groups.* Z. Kristallogr., Supplement Issue No. 1.

Fedorov, E. S. (1895). *Theorie der Kristallstruktur. Einleitung. Regelmässige Punktsysteme (mit übersichtlicher graphischer Darstellung).* Z. Kristallogr. **24**, 209–252, Tafel V, VI. [English translation by D. & K. Harker (1971). *Symmetry of crystals*, esp. pp. 206–213. Am. Crystallogr. Assoc., ACA Monograph No. 7.]

Fischer, W., Burzlaff, H., Hellner, E. & Donnay, J. D. H. (1973). *Space groups and lattice complexes.* NBS Monograph No. 134. Washington, DC: National Bureau of Standards.

Friedel, G. (1926). *Leçons de cristallographie.* Nancy/Paris/Strasbourg: Berger-Levrault. [Reprinted: Paris: Blanchard (1964).]

Heesch, H. (1929). *Zur systematischen Strukturtheorie. II.* Z. Kristallogr. **72**, 177–201.

Hilton, H. (1903). *Mathematical crystallography.* Oxford: Clarendon Press. [Reprint: New York: Dover (1963).]

Internationale Tabellen zur Bestimmung von Kristallstrukturen (1935). 1. Band, edited by C. Hermann. Berlin: Borntraeger. [Revised edition: Ann Arbor: Edwards (1944). Abbreviated as *IT* (1935).]

International Tables for Crystallography (2004). Vol. A1, edited by H. Wondratschek & U. Müller. Dordrecht: Kluwer Academic Publishers.

International Tables for Crystallography (2004). Vol. C, 3rd ed., edited by A. J. C. Wilson & E. Prince. Dordrecht: Kluwer Academic Publishers.

International Tables for X-ray Crystallography (1952). Vol. I, edited by N. F. M. Henry & K. Lonsdale. Birmingham: Kynoch Press. [Revised editions: 1965, 1969 and 1977. Abbreviated as *IT* (1952).]

Koch, E. & Fischer, W. (1974). *Zur Bestimmung asymmetrischer Einheiten kubischer Raumgruppen mit Hilfe von Wirkungsbereichen.* Acta Cryst. A**30**, 490–496.

Langlet, G. A. (1972). *FIGATOM: a new graphic program for stereographic crystal structure illustrations.* J. Appl. Cryst. **5**, 66–71.

Niggli, P. (1919). *Geometrische Kristallographie des Diskontinuums.* Leipzig: Borntraeger. [Reprint: Wiesbaden: Sändig (1973).]

Parthé, E., Gelato, L. M. & Chabot, B. (1988). *Structure description ambiguity depending upon which edition of International Tables for (X-ray) Crystallography is used.* Acta Cryst. A**44**, 999–1002.

Sadanaga, R., Takeuchi, Y. & Morimoto, N. (1978). *Complex structures of minerals.* Recent Prog. Nat. Sci. Jpn, **3**, 141–206, esp. pp. 149–151.

Schiebold, E. (1929). *Über eine neue Herleitung und Nomenklatur der 230 kristallographischen Raumgruppen mit Atlas der 230 Raumgruppen-Projektionen.* Text, Atlas. In *Abhandlungen der Mathematisch-Physikalischen Klasse der Sächsischen Akademie der Wissenschaften*, Band 40, Heft 5. Leipzig: Hirzel.

Templeton, D. H. (1956). *Systematic absences corresponding to false symmetry.* Acta Cryst. **9**, 199–200.

3. DETERMINATION OF SPACE GROUPS

By A. Looijenga-Vos and M. J. Buerger

3.1. Space-group determination and diffraction symbols

BY A. LOOIJENGA-VOS AND M. J. BUERGER

3.1.1. Introduction

In this chapter, the determination of space groups from the Laue symmetry and the reflection conditions, as obtained from diffraction patterns, is discussed. Apart from Section 3.1.6.5, where differences between reflections hkl and $\overline{hk}\,\overline{l}$ due to anomalous dispersion are discussed, it is assumed that Friedel's rule holds, i.e. that $|F(hkl)|^2 = |F(\overline{hk}\,\overline{l})|^2$. This implies that the reciprocal lattice weighted by $|F(hkl)|^2$ has an inversion centre, even if this is not the case for the crystal under consideration. Accordingly, the symmetry of the weighted reciprocal lattice belongs, as was discovered by Friedel (1913), to one of the eleven Laue classes of Table 3.1.2.1. As described in Section 3.1.5, Laue class plus reflection conditions in most cases do not uniquely specify the space group. Methods that help to overcome these ambiguities, especially with respect to the presence or absence of an inversion centre in the crystal, are summarized in Section 3.1.6.

3.1.2. Laue class and cell

Space-group determination starts with the assignment of the *Laue class* to the weighted reciprocal lattice and the determination of the *cell geometry*. The conventional cell (except for the case of a primitive rhombohedral cell) is chosen such that the basis vectors coincide as much as possible with directions of highest symmetry (cf. Chapters 2.1 and 9.1).

The axial system should be taken right-handed. For the different crystal systems, the symmetry directions (*blickrichtungen*) are listed in Table 2.2.4.1. The symmetry directions and the convention that, within the above restrictions, the cell should be taken as small as possible determine the axes and their labels uniquely for crystal systems with symmetry higher than orthorhombic. For orthorhombic crystals, three directions are fixed by symmetry, but any of the

three may be called a, b or c. For monoclinic crystals, there is one unique direction. It has to be decided whether this direction is called b, c or a. If there are no special reasons (physical properties, relations with other structures) to decide otherwise, the standard choice b is preferred. For triclinic crystals, usually the reduced cell is taken (cf. Chapter 9.2), but the labelling of the axes remains a matter of choice, as in the orthorhombic system.

If the lattice type turns out to be centred, which reveals itself by systematic absences in the general reflections hkl (Section 2.2.13), examination should be made to see whether the smallest cell has been selected, within the conventions appropriate to the crystal system. This is necessary since Table 3.1.4.1 for space-group determination is based on such a selection of the cell. Note, however, that for rhombohedral space groups two cells are considered, the triple hexagonal cell and the primitive rhombohedral cell.

The Laue class determines the crystal system. This is listed in Table 3.1.2.1. Note the conditions imposed on the lengths and the directions of the cell axes as well as the fact that there are crystal systems to which two Laue classes belong.

3.1.3. Reflection conditions and diffraction symbol

In Section 2.2.13, it has been shown that 'extinctions' (sets of reflections that are systematically absent) point to the presence of a centred cell or the presence of symmetry elements with glide or screw components. Reflection conditions and Laue class together are expressed by the *Diffraction symbol*, introduced by Buerger (1935, 1942, 1969); it consists of the Laue-class symbol, followed by the extinction symbol representing the observed reflection conditions. Donnay & Harker (1940) have used the concept of extinctions under the name of 'morphological aspect' (or aspect for short) in their studies of crystal habit (cf. *Crystal Data*, 1972). Although the concept of aspect applies to diffraction as well as to morphology (Donnay & Kennard, 1964), for the present tables the expression 'extinction symbol' has been chosen because of the morphological connotation of the word aspect.

The *Extinction symbols* are arranged as follows. First, a capital letter is given representing the centring type of the cell (Section 1.2.1). Thereafter, the reflection conditions for the successive symmetry directions are symbolized. Symmetry directions not having reflection conditions are represented by a dash. A symmetry direction with reflection conditions is represented by the symbol for the corresponding glide plane and/or screw axis. The symbols applied are the same as those used in the Hermann–Mauguin space-group symbols (Section 1.3.1). If a symmetry direction has more than one kind of glide plane, for the diffraction symbol the same letter is used as in the corresponding space-group symbol. An exception is made for some centred orthorhombic space groups where *two* glide-plane symbols are given (between parentheses) for one of the symmetry directions, in order to stress the relation between the diffraction symbol and the symbols of the 'possible space groups'. For the various orthorhombic settings, treated in Table 3.1.4.1, the top lines of the two-line space-group symbols in Table 4.3.2.1 are used. In the monoclinic system, dummy numbers '1' are inserted for two directions even though they are not symmetry directions, to bring out the differences between the diffraction symbols for the b, c and a settings.

Table 3.1.2.1. *Laue classes and crystal systems*

Laue class	Crystal system	Conditions imposed on cell geometry
$\overline{1}$	Triclinic	None
$2/m$	Monoclinic	$\alpha = \gamma = 90°$ (b unique)
		$\alpha = \beta = 90°$ (c unique)
mmm	Orthorhombic	$\alpha = \beta = \gamma = 90°$
$4/m$ $4/mmm$	Tetragonal	$a = b; \alpha = \beta = \gamma = 90°$
$\overline{3}$	Trigonal	$a = b; \alpha = \beta = 90°; \gamma = 120°$ (hexagonal axes)
$\overline{3}m$		$a = b = c; \alpha = \beta = \gamma$ (rhombohedral axes)
$6/m$ $6/mmm$	Hexagonal	$a = b; \alpha = \beta = 90°; \gamma = 120°$
$m\overline{3}$ $m\overline{3}m$	Cubic	$a = b = c; \alpha = \beta = \gamma = 90°$

44

Example

Laue class: $12/m1$

Reflection conditions:

$hkl : h + k = 2n$;

$h0l : h, l = 2n$; $0kl : k = 2n$; $hk0 : h + k = 2n$;

$h00 : h = 2n$; $0k0 : k = 2n$; $00l : l = 2n$.

As there are both c and n glide planes perpendicular to b, the diffraction symbol may be given as $1\,2/m\,1C1c1$ or as $1\,2/m\,1C1n1$. In analogy to the symbols of the possible space groups, $C1c1$ (9) and $C1\,2/c\,1$ (15), the diffraction symbol is called $1\,2/m\,1C1c1$.

For another cell choice, the reflection conditions are:

$hkl : k + l = 2n$;

$h0l : h, l = 2n$; $0kl : k + l = 2n$; $hk0 : k = 2n$;

$h00 : h = 2n$; $0k0 : k = 2n$; $00l : l = 2n$.

For this second cell choice, the glide planes perpendicular to b are n and a. The diffraction symbol is given as $1\,2/m\,1A1n1$, in analogy to the symbols $A1n1$ (9) and $A1\,2/n\,1$ (15) adopted for the possible space groups.

3.1.4. Deduction of possible space groups

Reflection conditions, diffraction symbols, and possible space groups are listed in Table 3.1.4.1. For each crystal system, a different table is provided. The monoclinic system contains different entries for the settings with b, c and a unique. For monoclinic and orthorhombic crystals, all possible settings and cell choices are treated. In contradistinction to Table 4.3.2.1, which lists the space-group symbols for different settings and cell choices in a systematic way, the present table is designed with the aim to make space-group determination as easy as possible.

The left-hand side of the table contains the *Reflection conditions*. Conditions of the type $h = 2n$ or $h + k = 2n$ are abbreviated as h or $h + k$. Conditions like $h = 2n, k = 2n, h + k = 2n$ are quoted as h, k; in this case, the condition $h + k = 2n$ is not listed as it follows directly from $h = 2n, k = 2n$. Conditions with $l = 3n$, $l = 4n$, $l = 6n$ or more complicated expressions are listed explicitly.

From *left* to *right*, the table contains the integral, zonal and serial conditions. From *top* to *bottom*, the entries are ordered such that left columns are kept empty as long as possible. The leftmost column that contains an entry is considered as the 'leading column'. In this column, entries are listed according to increasing complexity. This also holds for the subsequent columns within the restrictions imposed by previous columns on the left. The make-up of the table is such that observed reflection conditions should be matched against the table by considering, within each crystal system, the columns from left to right.

The centre column contains the *Extinction symbol*. To obtain the complete diffraction symbol, the Laue-class symbol has to be added in front of it. Be sure that the correct Laue-class symbol is used if the crystal system contains two Laue classes. Particular care is needed for Laue class $\bar{3}m$ in the trigonal system, because there are two possible orientations of this Laue symmetry with respect to the crystal lattice, $\bar{3}m1$ and $\bar{3}1m$. The correct orientation can be obtained directly from the diffraction record.

The right-hand side of the table gives the *Possible space groups* which obey the reflection conditions. For crystal systems with two Laue classes, a subdivision is made according to the Laue symmetry. The entries in each Laue class are ordered according to their point groups. All space groups that match both the reflection

conditions and the Laue symmetry, found in a diffraction experiment, are possible space groups of the crystal.

The space groups are given by their short Hermann–Mauguin symbols, followed by their number between parentheses, except for the monoclinic system, where full symbols are given (*cf.* Section 2.2.4). In the monoclinic and orthorhombic sections of Table 3.1.4.1, which contain entries for the different settings and cell choices, the 'standard' space-group symbols (*cf.* Table 4.3.2.1) are printed in bold face. Only these standard representations are treated in full in the space-group tables.

Example

The diffraction pattern of a compound has Laue class *mmm*. The crystal system is thus orthorhombic. The diffraction spots are indexed such that the reflection conditions are $0kl : l = 2n$; $h0l : h + l = 2n$; $h00 : h = 2n$; $00l : l = 2n$. Table 3.1.4.1 shows that the diffraction symbol is *mmmPcn–*. Possible space groups are *Pcn*2 (30) and *Pcnm* (53). For neither space group does the axial choice correspond to that of the standard setting. For No. 30, the standard symbol is *Pnc*2, for No. 53 it is *Pmna*. The transformation from the basis vectors $\mathbf{a}_e, \mathbf{b}_e, \mathbf{c}_e$, used in the experiment, to the basis vectors $\mathbf{a}_s, \mathbf{b}_s, \mathbf{c}_s$ of the standard setting is given by $\mathbf{a}_s = \mathbf{b}_e, \mathbf{b}_s = -\mathbf{a}_e$ for No. 30 and by $\mathbf{a}_s = \mathbf{c}_e, \mathbf{c}_s = -\mathbf{a}_e$ for No. 53.

Possible pitfalls

Errors in the space-group determination may occur because of several reasons.

(1) *Twinning of the crystal*

Difficulties that may be encountered are shown by the following example. Say that a monoclinic crystal (b unique) with the angle β fortuitously equal to $\sim 90°$ is twinned according to (100). As this causes overlap of the reflections hkl and $\bar{h}kl$, the observed Laue symmetry is *mmm* rather than $2/m$. The same effect may occur within one crystal system. If, for instance, a crystal with Laue class $4/m$ is twinned according to (100) or (110), the Laue class $4/mmm$ is simulated (twinning by merohedry, *cf.* Catti & Ferraris, 1976, and Koch, 1999). Further examples are given by Buerger (1960). Errors due to twinning can often be detected from the fact that the observed reflection conditions do not match any of the diffraction symbols.

(2) *Incorrect determination of reflection conditions*

Either too many or too few conditions may be found. For serial reflections, the first case may arise if the structure is such that its projection on, say, the b direction shows pseudo-periodicity. If the pseudo-axis is b/p, with p an integer, the reflections $0k0$ with $k \neq p$ are very weak. If the exposure time is not long enough, they may be classified as unobserved which, incorrectly, would lead to the reflection condition $0k0 : k = p$. A similar situation may arise for zonal conditions, although in this case there is less danger of errors. Many more reflections are involved and the occurrence of pseudo-periodicity is less likely for two-dimensional than for one-dimensional projections.

For 'structural' or non-space-group absences, see Section 2.2.13.

The second case, too many observed reflections, may be due to multiple diffraction or to radiation impurity. A textbook description of multiple diffraction has been given by Lipson & Cochran (1966). A well known case of radiation impurity in X-ray diffraction is the contamination of a copper target with iron. On a photograph taken with the radiation from such a target, the iron radiation with $\lambda(\text{Fe}) \sim 5/4\lambda(\text{Cu})$ gives a reflection spot $4h,4k,4l$ at the position $5h,5k,5l$ for copper $[\lambda(\text{Cu}\,K\bar{\alpha}) = 1.5418$ Å, $\lambda(\text{Fe}\,K\bar{\alpha}) = 1.9373$ Å$]$. For reflections $0k0$, for instance, this may give rise to reflected intensity at the copper 050 position so that, incorrectly, the condition $0k0 : k = 2n$ may be excluded.

3. DETERMINATION OF SPACE GROUPS

Table 3.1.4.1. *Reflection conditions, diffraction symbols and possible space groups*

TRICLINIC. Laue class $\bar{1}$

Reflection conditions	Extinction symbol	Point group	
		1	$\bar{1}$
None	$P-$	$P1(1)$	$P\bar{1}$ (2)

MONOCLINIC, Laue class $2/m$

Unique axis b				Laue class $1\,2/m\,1$			
Reflection conditions				**Point group**			
hkl $0kl\ hk0$	$h0l$ $h00\ 00l$	$0k0$	Extinction symbol	2	m	$2/m$	
			$P1-1$	**$P121$** (3)	**$P1m1$** (6)	**$P1\,2/m\,1$** (10)	
		k	$P12_11$	**$P12_11$** (4)		**$P1\,2_1/m\,1$** (11)	
	h		$P1a1$		$P1a1$ (7)	$P1\,2/a\,1$ (13)	
	h	k	$P1\,2_1/a\,1$			$P1\,2_1/a\,1$ (14)	
	l		$P1c1$		**$P1c1$** (7)	**$P1\,2/c\,1$** (13)	
	l	k	$P1\,2_1/c\,1$			**$P1\,2_1/c\,1$** (14)	
	$h+l$		$P1n1$		$P1n1$ (7)	$P1\,2/n\,1$ (13)	
	$h+l$	k	$P1\,2_1/n\,1$			$P1\,2_1/n\,1$ (14)	
$h+k$	h	k	$C1-1$	**$C121$** (5)	**$C1m1$** (8)	**$C1\,2/m\,1$** (12)	
$h+k$	h,l	k	$C1c1$		$C1c1$ (9)	$C1\,2/c\,1$ (15)	
$k+l$	l	k	$A1-1$	$A121$ (5)	$A1m1$ (8)	$A1\,2/m\,1$ (12)	
$k+l$	h,l	k	$A1n1$		$A1n1$ (9)	$A1\,2/n\,1$ (15)	
$h+k+l$	$h+l$	k	$I1-1$	$I121$ (5)	$I1m1$ (8)	$I1\,2/m\,1$ (12)	
$h+k+l$	h,l	k	$I1a1$		$I1a1$ (9)	$I1\,2/a\,1$ (15)	

Unique axis c				Laue class $1\,1\,2/m$			
Reflection conditions				**Point group**			
hkl $0kl\ h0l$	$hk0$ $h00\ 0k0$	$00l$	Extinction symbol	2	m	$2/m$	
			$P11-$	$P112$ (3)	$P11m$ (6)	$P11\,2/m$ (10)	
		l	$P112_1$	$P112_1$ (4)		$P11\,2_1/m$ (11)	
	h		$P11a$		$P11a$ (7)	$P11\,2/a$ (13)	
	h	l	$P11\,2_1/a$			$P11\,2_1/a$ (14)	
	k		$P11b$		$P11b$ (7)	$P11\,2/b$ (13)	
	k	l	$P11\,2_1/b$			$P11\,2_1/b$ (14)	
	$h+k$		$P11n$		$P11n$ (7)	$P11\,2/n$ (13)	
	$h+k$	l	$P11\,2_1/n$			$P11\,2_1/n$ (14)	
$h+l$	h	l	$B11-$	$B112$ (5)	$B11m$ (8)	$B11\,2/m$ (12)	
$h+l$	h,k	l	$B11n$		$B11n$ (9)	$B11\,2/n$ (15)	
$k+l$	k	l	$A11-$	$A112$ (5)	$A11m$ (8)	$A11\,2/m$ (12)	
$k+l$	h,k	l	$A11a$		$A11a$ (9)	$A11\,2/a$ (15)	
$h+k+l$	$h+k$	l	$I11-$	$I112$ (5)	$I11m$ (8)	$I11\,2/m$ (12)	
$h+k+l$	h,k	l	$I11b$		$I11b$ (9)	$I11\,2/b$ (15)	

(3) *Incorrect assignment of the Laue symmetry*

This may be caused by pseudo-symmetry or by 'diffraction enhancement'. A crystal with pseudo-symmetry shows small deviations from a certain symmetry, and careful inspection of the diffraction pattern is necessary to determine the correct Laue class. In the case of diffraction enhancement, the symmetry of the diffraction pattern is higher than the Laue symmetry of the crystal. Structure types showing this phenomenon are rare and have to fulfil specified conditions. For further discussions and references, see Perez-Mato & Iglesias (1977).

3.1.5. Diffraction symbols and possible space groups

Table 3.1.4.1 contains 219 extinction symbols which, when combined with the Laue classes, lead to 242 different diffraction symbols. If, however, for the monoclinic and orthorhombic systems

Table 3.1.4.1. *Reflection conditions, diffraction symbols and possible space groups* (*cont.*)

MONOCLINIC, Laue class $2/m$ (*cont.*)

Unique axis a					Laue class $2/m\,1\,1$			
Reflection conditions					Point group			
hkl $h0l\ hk0$	$0kl$ $0k0\ 00l$		$h00$	Extinction symbol	2	m	$2/m$	
				P–11	$P211$ (3)	$Pm11$ (6)	$P2/m\,11$ (10)	
			h	$P2_1 11$	$P2_1 11$ (4)		$P2_1/m\,11$ (11)	
	k			$Pb11$		$Pb11$ (7)	$P2/b\,11$ (13)	
	k		h	$P2_1/b\,11$			$P2_1/b\,11$ (14)	
	l			$Pc11$		$Pc11$ (7)	$P2/c\,11$ (13)	
	l		h	$P2_1/c\,11$			$P2_1/c\,11$ (14)	
	$k+l$			$Pn11$		$Pn11$ (7)	$P2/n\,11$ (13)	
	$k+l$		h	$P2_1/n\,11$			$P2_1/n\,11$ (14)	
$h+k$	k		h	C–11	$C211$ (5)	$Cm11$ (8)	$C2/m\,11$ (12)	
$h+k$	k,l		h	$Cn11$		$Cn11$ (9)	$C2/n\,11$ (15)	
$h+l$	l		h	B–11	$B211$ (5)	$Bm11$ (8)	$B2/m\,11$ (12)	
$h+l$	k,l		h	$Bb11$		$Bb11$ (9)	$B2/b\,11$ (15)	
$h+k+l$	$k+l$		h	I–11	$I211$ (5)	$Im11$ (8)	$I2/m\,11$ (12)	
$h+k+l$	k,l		h	$Ic11$		$Ic11$ (9)	$I2/c\,11$ (15)	

ORTHORHOMBIC, Laue class mmm ($2/m\,2/m\,2/m$)
In this table, the symbol e in the space-group symbol represents the two glide planes given between parentheses in the corresponding extinction symbol. Only for one of the two cases does a bold printed symbol correspond with the standard symbol.

Reflection conditions								Laue class mmm ($2/m\,2/m\,2/m$)			
									Point group		
hkl	$0kl$	$h0l$	$hk0$	$h00$	$0k0$	$00l$	Extinction symbol	222	mm2 m2m 2mm	mmm	
							$P---$	$P222$ (16)	$Pmm2$ (25) $Pm2m$ (25) $P2mm$ (25)	$Pmmm$ (47)	
						l	$P--2_1$	$P222_1$ (17)			
					k		$P-2_1-$	$P22_12$ (17)			
					k	l	$P-2_12_1$	$P22_12_1$ (18)			
				h			$P2_1--$	$P2_122$ (17)			
				h		l	$P2_1-2_1$	$P2_122_1$ (18)			
				h	k		$P2_12_1-$	$P2_12_12$ (18)			
				h	k	l	$P2_12_12_1$	$P2_12_12_1$ (19)			
			h	h			$P--a$		$Pm2a$ (28) $P2_1ma$ (26)	$Pmma$ (51)	
		k			k		$P--b$		$Pm2_1b$ (26) $P2mb$ (28)	$Pmmb$ (51)	
		$h+k$		h	k		$P--n$		$Pm2_1n$ (31) $P2_1mn$ (31)	$Pmmn$ (59)	
	h			h			$P-a-$		$Pma2$ (28) $P2_1am$ (26)	$Pmam$ (51)	
	h	h		h			$P-aa$		$P2aa$ (27)	$Pmaa$ (49)	
	h	k		h	k		$P-ab$		$P2_1ab$ (29)	$Pmab$ (57)	
	h	$h+k$		h	k		$P-an$		$P2an$ (30)	$Pman$ (53)	
	l					l	$P-c-$		$Pmc2_1$ (26) $P2cm$ (28)	$Pmcm$ (51)	
	l	h		h		l	$P-ca$		$P2_1ca$ (29)	$Pmca$ (57)	
	l	k			k	l	$P-cb$		$P2cb$ (32)	$Pmcb$ (55)	
	l	$h+k$		h	k	l	$P-cn$		$P2_1cn$ (33)	$Pmcn$ (62)	

Table 3.1.4.1. *Reflection conditions, diffraction symbols and possible space groups* (cont.)

ORTHORHOMBIC, Laue class *mmm* (2/m 2/m 2/m) (cont.)

Reflection conditions								Laue class *mmm* (2/m 2/m 2/m)		
								Point group		
									mm2 m2m 2mm	
							Extinction symbol			
hkl	0*kl*	*h0l*	*hk*0	*h*00	0*k*0	00*l*	symbol	222		*mmm*
		$h+l$		h		l	P–n–		**Pmn2₁** (31)	
									P2₁nm (31)	Pmnm (59)
		$h+l$	h	h		l	P–na		P2na (30)	**Pmna** (53)
		$h+l$	k	h	k	l	P–nb		P2₁nb (33)	Pmnb (62)
		$h+l$	$h+k$	h	k	l	P–nn		P2nn (34)	Pmnn (58)
	k				k		Pb– –		Pbm2 (28)	
									Pb2₁m (26)	Pbmm (51)
	k		h	h	k		Pb–a		Pb2₁a (29)	Pbma (57)
	k		k		k		Pb–b		Pb2b (27)	Pbmb (49)
	k		$h+k$	h	k		Pb–n		Pb2n (30)	Pbmn (53)
	k	h		h	k		Pba–		**Pba2** (32)	**Pbam** (55)
	k	h	h	h	k		Pbaa			Pbaa (54)
	k	h	k	h	k		Pbab			Pbab (54)
	k	h	$h+k$	h	k		Pban			**Pban** (50)
	k	l			k	l	Pbc–		Pbc2₁ (29)	**Pbcm** (57)
	k	l	h	h	k	l	Pbca			**Pbca** (61)
	k	l	k		k	l	Pbcb			Pbcb (54)
	k	l	$h+k$	h	k	l	Pbcn			**Pbcn** (60)
	k	$h+l$		h	k	l	Pbn–		Pbn2₁ (33)	Pbnm (62)
	k	$h+l$	h	h	k	l	Pbna			Pbna (60)
	k	$h+l$	k	h	k	l	Pbnb			Pbnb (56)
	k	$h+l$	$h+k$	h	k	l	Pbnn			Pbnn (52)
	l					l	Pc– –		Pcm2₁ (26)	
									Pc2m (28)	Pcmm (51)
	l		h	h		l	Pc–a		Pc2a (32)	Pcma (55)
	l		k		k	l	Pc–b		Pc2₁b (29)	Pcmb (57)
	l		$h+k$	h	k	l	Pc–n		Pc2₁n (33)	Pcmn (62)
	l	h		h		l	Pca–		**Pca2₁** (29)	Pcam (57)
	l	h	h	h		l	Pcaa			Pcaa (54)
	l	h	k	h	k	l	Pcab			Pcab (61)
	l	h	$h+k$	h	k	l	Pcan			Pcan (60)
	l	l				l	Pcc–		**Pcc2** (27)	**Pccm** (49)
	l	l	h	h		l	Pcca			**Pcca** (54)
	l	l	k		k	l	Pccb			Pccb (54)
	l	l	$h+k$	h	k	l	Pccn			**Pccn** (56)
	l	$h+l$		h		l	Pcn –		Pcn2 (30)	Pcnm (53)
	l	$h+l$	h	h		l	Pcna			Pcna (50)
	l	$h+l$	k	h	k	l	Pcnb			Pcnb (60)
	l	$h+l$	$h+k$	h	k	l	Pcnn			Pcnn (52)
	$k+l$				k	l	Pn – –		Pnm2₁ (31)	Pnmm (59)
									Pn2₁m (31)	
	$k+l$		h	h	k	l	Pn–a		Pn2₁a (33)	**Pnma** (62)
	$k+l$		k		k	l	Pn–b		Pn2b (30)	Pnmb (53)
	$k+l$		$h+k$	h	k	l	Pn–n		Pn2n (34)	Pnmn (58)
	$k+l$	h		h	k	l	Pna –		**Pna2₁** (33)	Pnam (62)
	$k+l$	h	h	h	k	l	Pnaa			Pnaa (56)
	$k+l$	h	k	h	k	l	Pnab			Pnab (60)
	$k+l$	h	$h+k$	h	k	l	Pnan			Pnan (52)
	$k+l$	l			k	l	Pnc –		Pnc2 (30)	Pncm (53)
	$k+l$	l	h	h	k	l	Pnca			Pnca (60)

Table 3.1.4.1. *Reflection conditions, diffraction symbols and possible space groups* (*cont.*)

ORTHORHOMBIC, Laue class mmm ($2/m\ 2/m\ 2/m$) (*cont.*)

Reflection conditions								Laue class mmm ($2/m\ 2/m\ 2/m$)		
								Point group		
hkl	$0kl$	$h0l$	$hk0$	$h00$	$0k0$	$00l$	Extinction symbol	222	$mm2$ $m2m$ $2mm$	mmm
	$k+l$	l	k		k	l	$Pncb$			$Pncb$ (50)
	$k+l$	l	$h+k$	h	k	l	$Pncn$			$Pncn$ (52)
	$k+l$	$h+l$		h	k	l	$Pnn-$		$\mathbf{Pnn2}$ (34)	$\mathbf{Pnnm}$ (58)
	$k+l$	$h+l$	h	h	k	l	$Pnna$			$\mathbf{Pnna}$ (52)
	$k+l$	$h+l$	k	h	k	l	$Pnnb$			$Pnnb$ (52)
	$k+l$	$h+l$	$h+k$	h	k	l	$Pnnn$			$\mathbf{Pnnn}$ (48)
$h+k$	k	h	$h+k$	h	k		$C---$	$\mathbf{C222}$ (21)	$\mathbf{Cmm2}$ (35) $Cm2m$ (38) $C2mm$ (38)	$\mathbf{Cmmm}$ (65)
$h+k$	k	h	$h+k$	h	k	l	$C--2_1$	$\mathbf{C222_1}$ (20)		
$h+k$	k	h	h,k	h	k		$C--(ab)$		$Cm2e$ (39) $C2me$ (39)	$\mathbf{Cmme}$ (67)
$h+k$	k	h,l	$h+k$	h	k	l	$C-c-$		$\mathbf{Cmc2_1}$ (36) $C2cm$ (40)	$\mathbf{Cmcm}$ (63)
$h+k$	k	h,l	h,k	h	k	l	$C-c(ab)$		$C2ce$ (41)	$\mathbf{Cmce}$ (64)
$h+k$	k,l	h	$h+k$	h	k	l	$Cc--$		$Ccm2_1$ (36) $Cc2m$ (40)	$Ccmm$ (63)
$h+k$	k,l	h	h,k	h	k	l	$Cc-(ab)$		$Cc2e$ (41)	$Ccme$ (64)
$h+k$	k,l	h,l	$h+k$	h	k	l	$Ccc-$		$\mathbf{Ccc2}$ (37)	$\mathbf{Cccm}$ (66)
$h+k$	k,l	h,l	h,k	h	k	l	$Ccc(ab)$			$\mathbf{Ccce}$ (68)
$h+l$	l	$h+l$	h	h		l	$B---$	$B222$ (21)	$Bmm2$ (38) $Bm2m$ (35) $B2mm$ (38)	$Bmmm$ (65)
$h+l$	l	$h+l$	h	h	k	l	$B-2_1-$	$B22_12$ (20)		
$h+l$	l	$h+l$	h,k	h	k	l	$B--b$		$Bm2_1b$ (36) $B2mb$ (40)	$Bmmb$ (63)
$h+l$	l	h,l	h	h		l	$B-(ac)-$		$Bme2$ (39) $B2em$ (39)	$Bmem$ (67)
$h+l$	l	h,l	h,k	h	k	l	$B-(ac)b$		$B2eb$ (41)	$Bmeb$ (64)
$h+l$	k,l	$h+l$	h	h	k	l	$Bb--$		$Bbm2$ (40) $Bb2_1m$ (36)	$Bbmm$ (63)
$h+l$	k,l	$h+l$	h,k	h	k	l	$Bb-b$		$Bb2b$ (37)	$Bbmb$ (66)
$h+l$	k,l	h,l	h	h	k	l	$Bb(ac)-$		$Bbe2$ (41)	$Bbem$ (64)
$h+l$	k,l	h,l	h,k	h	k	l	$Bb(ac)b$			$Bbeb$ (68)
$k+l$	$k+l$	l	k		k	l	$A---$	$A222$ (21)	$\mathbf{Amm2}$ (38) $Am2m$ (38) $A2mm$ (35)	$Ammm$ (65)
$k+l$	$k+l$	l	k	h	k	l	$A2_1--$	$A2_122$ (20)		
$k+l$	$k+l$	l	h,k	h	k	l	$A--a$		$\mathbf{Am2a}$ (40) $A2_1ma$ (36)	$Amma$ (63)
$k+l$	$k+l$	h,l	k	h	k	l	$A-a-$		$\mathbf{Ama2}$ (40) $A2_1am$ (36)	$Amam$ (63)
$k+l$	$k+l$	h,l	h,k	h	k	l	$A-aa$		$A2aa$ (37)	$Amaa$ (66)
$k+l$	k,l	l	k		k	l	$A(bc)--$		$\mathbf{Aem2}$ (39) $Ae2m$ (39)	$Aemm$ (67)
$k+l$	k,l	l	h,k	h	k	l	$A(bc)-a$		$Ae2a$ (41)	$Aema$ (64)
$k+l$	k,l	h,l	k	h	k	l	$A(bc)a-$		$\mathbf{Aea2}$ (41)	$Aeam$ (64)
$k+l$	k,l	h,l	h,k	h	k	l	$A(bc)aa$			$Aeaa$ (68)
$h+k+l$	$k+l$	$h+l$	$h+k$	h	k	l	$I---$	$\begin{bmatrix}\mathbf{I222}\ (23)\\ \mathbf{I2_12_12_1}(24)\end{bmatrix}_*$	$\mathbf{Imm2}$ (44) $Im2m$ (44)	$\mathbf{Immm}$ (71)

49

Table 3.1.4.1. *Reflection conditions, diffraction symbols and possible space groups* (cont.)

ORTHORHOMBIC, Laue class *mmm* (2/*m* 2/*m* 2/*m*) (cont.)

Reflection conditions								Laue class *mmm* (2/*m* 2/*m* 2/*m*)		
									Point group	
hkl	0*kl*	*h*0*l*	*hk*0	*h*00	0*k*0	00*l*	Extinction symbol	222	mm2 m2m 2mm	mmm
$h+k+l$	$k+l$	$h+l$	h, k	h	k	l	$I--(ab)$		I2mm (44) Im2a (46) I2mb (46)	**Imma** (74) Immb (74)
$h+k+l$	$k+l$	h, l	$h+k$	h	k	l	$I-(ac)-$		**Ima2** (46) I2cm (46)	Imam (74) Imcm (74)
$h+k+l$	$k+l$	h, l	h, k	h	k	l	$I-cb$		I2cb (45)	Imcb (72)
$h+k+l$	k, l	$h+l$	$h+k$	h	k	l	$I(bc)--$		Iem2 (46) Ie2m (46)	Iemm (74)
$h+k+l$	k, l	$h+l$	h, k	h	k	l	$Ic-a$		Ic2a (45)	Icma (72)
$h+k+l$	k, l	h, l	$h+k$	h	k	l	$Iba-$		**Iba2** (45)	**Ibam** (72)
$h+k+l$	k, l	h, l	h, k	h	k	l	$Ibca$			**Ibca** (73) Icab (73)
$h+k, h+l, k+l$	k, l	h, l	h, k	h	k	l	$F---$	**F222** (22)	**Fmm2** (42) Fm2m (42) F2mm (42)	**Fmmm** (69)
$h+k, h+l, k+l$	k, l	$h+l=4n; h, l$	$h+k=4n; h, k$	$h=4n$	$k=4n$	$l=4n$	$F-dd$		F2dd (43)	
$h+k, h+l, k+l$	$k+l=4n; k, l$	h, l	$h+k=4n; h, k$	$h=4n$	$k=4n$	$l=4n$	$Fd-d$		Fd2d (43)	
$h+k, h+l, k+l$	$k+l=4n; k, l$	$h+l=4n; h, l$	h, k	$h=4n$	$k=4n$	$l=4n$	$Fdd-$		**Fdd2** (43)	
$h+k, h+l, k+l$	$k+l=4n; k, l$	$h+l=4n; h, l$	$h+k=4n; h, k$	$h=4n$	$k=4n$	$l=4n$	$Fddd$			**Fddd** (70)

* Pair of space groups with common point group and symmetry elements but differing in the relative location of these elements.

TETRAGONAL, Laue classes 4/*m* and 4/*mmm*

Reflection conditions							Extinction symbol	Laue class						
								4/*m*			4/*mmm* (4/*m* 2/*m* 2/*m*)			
								Point group						
hkl	*hk*0	0*kl*	*hhl*	00*l*	0*k*0	*hh*0		4	$\bar{4}$	4/*m*	422	4mm	$\bar{4}2m$ $\bar{4}m2$	4/*mmm*
							$P---$	P4 (75)	P$\bar{4}$ (81)	P4/m (83)	P422 (89)	P4mm (99)	P$\bar{4}2m$ (111) P$\bar{4}m2$ (115)	P4/mmm (123)
					k		$P-2_1-$				P42₁2 (90)		P$\bar{4}2_1m$ (113)	
			l				$P4_2--$	P4₂ (77)		P4₂/m (84)	P4₂22 (93)			
			l		k		$P4_22_1-$				P4₂2₁2 (94)			
			$l=4n$				$P4_1--$	$\begin{Bmatrix} P4_1\ (76) \\ P4_3\ (78) \end{Bmatrix}$†			$\begin{Bmatrix} P4_122\ (91) \\ P4_322\ (95) \end{Bmatrix}$†			
			$l=4n$		k		$P4_12_1-$				$\begin{Bmatrix} P4_12_12\ (92) \\ P4_32_12\ (96) \end{Bmatrix}$†			
		l	l				$P--c$					P4₂mc (105)	P$\bar{4}2c$ (112)	P4₂/mmc (131)
		l	l		k		$P-2_1c$						P$\bar{4}2_1c$ (114)	
	k				k		$P-b-$					P4bm (100)	P$\bar{4}b2$ (117)	P4/mbm (127)
	k	l		l	k		$P-bc$					P4₂bc (106)		P4₂/mbc (135)
	l			l			$P-c-$					P4₂cm (101)	P$\bar{4}c2$ (116)	P4₂/mcm (132)
	l	l		l			$P-cc$					P4cc (103)		P4/mcc (124)
	$k+l$			l	k		$P-n-$					P4₂nm (102)	P$\bar{4}n2$ (118)	P4₂/mnm (136)
	$k+l$	l		l	k		$P-nc$					P4nc (104)		P4/mnc (128)
$h+k$					k		$Pn--$			P4/n (85)				P4/nmm (129)
$h+k$			l		k		$P4_2/n--$			P4₂/n (86)				
$h+k$		l	l		k		$Pn-c$							P4₂/nmc (137)

Table 3.1.4.1. *Reflection conditions, diffraction symbols and possible space groups* (*cont.*)

TETRAGONAL, Laue classes $4/m$ and $4/mmm$ (*cont.*)

	Reflection conditions								Laue class						
									$4/m$			$4/mmm$ $(4/m\,2/m\,2/m)$			
									Point group						
hkl	$hk0$	$0kl$	hhl	$00l$	$0k0$	$hh0$	Extinction symbol		4	$\bar{4}$	$4/m$	422	$4mm$	$\bar{4}2m\ \bar{4}m2$	$4/mmm$
	$h+k$	k			k		$Pnb-$								$P4/nbm$ (125)
	$h+k$	k	l	l	k		$Pnbc$								$P4_2/nbc$ (133)
	$h+k$	l		l	k		$Pnc-$								$P4_2/ncm$ (138)
	$h+k$	l	l	l	k		$Pncc$								$P4/ncc$ (130)
	$h+k$	$k+l$		l	k		$Pnn-$								$P4_2/nnm$ (134)
	$h+k$	$k+l$	l	l	k		$Pnnc$								$P4/nnc$ (126)
$h+k+l$	$h+k$	$k+l$	l	l	k		$I---$		$I4$ (79)	$I\bar{4}$ (82)	$I4/m$ (87)	$I422$ (97)	$I4mm$ (107)	$I\bar{4}2m$ (121) $I\bar{4}m2$ (119)	$I4/mmm$ (139)
$h+k+l$	$h+k$	$k+l$	l	$l=4n$	k		$I4_1--$		$I4_1$ (80)			$I4_122$ (98)			
$h+k+l$	$h+k$	$k+l$	‡	$l=4n$	k	h	$I--d$						$I4_1md$ (109)	$I\bar{4}2d$ (122)	
$h+k+l$	$h+k$	k,l	l	l	k		$I-c-$						$I4cm$ (108)	$I\bar{4}c2$ (120)	$I4/mcm$ (140)
$h+k+l$	$h+k$	k,l	‡	$l=4n$	k	h	$I-cd$						$I4_1cd$ (110)		
$h+k+l$	h,k	$k+l$	l	$l=4n$	k		$I4_1/a--$				$I4_1/a$ (88)				
$h+k+l$	h,k	$k+l$	‡	$l=4n$	k	h	$Ia-d$								$I4_1/amd$ (141)
$h+k+l$	h,k	k,l	‡	$l=4n$	k	h	$Iacd$								$I4_1/acd$ (142)

† Pair of enantiomorphic space groups, *cf.* Section 3.1.5.
‡ Condition: $2h+l=4n$; l.

(as well as for the R space groups of the trigonal system), the different cell choices and settings of one space group are disregarded, 101 extinction symbols* and 122 diffraction symbols for the 230 space-group types result.

Only in 50 cases does a diffraction symbol uniquely identify just one space group, thus leaving 72 diffraction symbols that correspond to more than one space group. The 50 unique cases can be easily recognized in Table 3.1.4.1 because the line for the possible space groups in the particular Laue class contains just one entry.

The non-uniqueness of the space-group determination has two reasons:

(i) Friedel's rule, *i.e.* the effect that, with neglect of anomalous dispersion, the diffraction pattern contains an inversion centre, even if such a centre is not present in the crystal.

Example

A monoclinic crystal (with unique axis b) has the diffraction symbol $1\,2/m\,1P1c1$. Possible space groups are $P1c1$ (7) without an inversion centre, and $P12/c1$ (13) with an inversion centre. In both cases, the diffraction pattern has the Laue symmetry $1\,2/m\,1$.

One aspect of Friedel's rule is that the diffraction patterns are the same for two enantiomorphic space groups. Eleven diffraction symbols each correspond to a pair of enantiomorphic space groups.

In Table 3.1.4.1, such pairs are grouped between braces. Either of the two space groups may be chosen for structure solution. If due to anomalous scattering Friedel's rule does not hold, at the refinement stage of structure determination it may be possible to determine the absolute structure and consequently the correct space group from the enantiomorphic pair.

(ii) The occurrence of four space groups in two 'special' pairs, each pair belonging to the same point group: $I222$ (23) & $I2_12_12_1$ (24) and $I23$ (197) & $I2_13$ (199). The two space groups of each pair differ in the location of the symmetry elements with respect to each other. In Table 3.1.4.1, these two special pairs are given in square brackets.

3.1.6. Space-group determination by additional methods

3.1.6.1. *Chemical information*

In some cases, chemical information determines whether or not the space group is centrosymmetric. For instance, all proteins crystallize in noncentrosymmetric space groups as they are constituted of L-amino acids only. Less certain indications may be obtained by considering the number of molecules per cell and the possible space-group symmetry. For instance, if experiment shows that there are two molecules of formula $A_\alpha B_\beta$ per cell in either space group $P2_1$ or $P2_1/m$ and if the molecule $A_\alpha B_\beta$ cannot possibly have either a mirror plane or an inversion centre, then there is a strong indication that the correct space group is $P2_1$. Crystallization of $A_\alpha B_\beta$ in $P2_1/m$ with random disorder of the molecules cannot be excluded, however. In a similar way, multiplicities of Wyckoff positions and the number of formula units per cell may be used to distinguish between space groups.

---

* The increase from 97 (*IT*, 1952) to 101 extinction symbols is due to the separate treatment of the trigonal and hexagonal crystal systems in Table 3.1.4.1, in contradistinction to *IT* (1952), Table 4.4.3, where they were treated together. In *IT* (1969), diffraction symbols were listed by Laue classes and thus the number of extinction symbols is the same as that of diffraction symbols, namely 122.

Table 3.1.4.1. *Reflection conditions, diffraction symbols and possible space groups (cont.)*

TRIGONAL, Laue classes $\bar{3}$ and $\bar{3}m$

Reflection conditions					Laue class								
						$\bar{3}$		$\bar{3}m1\ (\bar{3}\ 2/m\ 1)$ $\bar{3}m$			$\bar{3}1m\ (\bar{3}\ 1\ 2/m)$		
Hexagonal axes						Point group							
$hkil$	$h\bar{h}0l$	$hh\overline{2h}l$	$000l$	Extinction symbol		3	$\bar{3}$	321 32	$3m1$ $3m$	$\bar{3}m1$ $\bar{3}m$	312	$31m$	$\bar{3}1m$
				$P---$		$P3$ (143)	$P\bar{3}$ (147)	$P321$ (150)	$P3m1$ (156)	$P\bar{3}m1$ (164)	$P312$ (149)	$P31m$ (157)	$P\bar{3}1m$ (162)
			$l=3n$	$P3_1--$		$\left\{\begin{array}{l}P3_1\,(144)\\P3_2\,(145)\end{array}\right\}§$		$\left\{\begin{array}{l}P3_121\,(152)\\P3_221\,(154)\end{array}\right\}§$			$\left\{\begin{array}{l}P3_112\,(151)\\P3_212\,(153)\end{array}\right\}§$		
		l	l	$P--c$								$P31c$ (159)	$P\bar{3}1c$ (163)
	l		l	$P-c-$					$P3c1$ (158)	$P\bar{3}c1$ (165)			
$-h+k+l=3n$	$h+l=3n$	$l=3n$	$l=3n$	$R(\text{obv})--$ ¶		$R3$ (146)	$R\bar{3}$ (148)	$R32$ (155)	$R3m$ (160)	$R\bar{3}m$ (166)			
$-h+k+l=3n$	$h+l=3n;\ l$	$l=3n$	$l=6n$	$R(\text{obv})-c$					$R3c$ (161)	$R\bar{3}c$ (167)			
$h-k+l=3n$	$-h+l=3n$	$l=3n$	$l=3n$	$R(\text{rev})--$		$R3$ (146)	$R\bar{3}$ (148)	$R32$ (155)	$R3m$ (160)	$R\bar{3}m$ (166)			
$h-k+l=3n$	$-h+l=3n;\ l$	$l=3n$	$l=6n$	$R(\text{rev})-c$					$R3c$ (161)	$R\bar{3}c$ (167)			
Rhombohedral axes			Extinction symbol			Point group							
hkl	hhl	hhh				3	$\bar{3}$	32	$3m$	$\bar{3}m$			
			$R--$			$R3$ (146)	$R\bar{3}$ (148)	$R32$ (155)	$R3m$ (160)	$R\bar{3}m$ (166)			
	l	h	$R-c$						$R3c$ (161)	$R\bar{3}c$ (167)			

§ Pair of enantiomorphic space groups; *cf.* Section 3.1.5.
¶ For obverse and reverse settings *cf.* Section 1.2.1. The obverse setting is standard in these tables.
The transformation reverse → obverse is given by $\mathbf{a}(\text{obv.}) = -\mathbf{a}(\text{rev.})$, $\mathbf{b}(\text{obv.}) = -\mathbf{b}(\text{rev.})$, $\mathbf{c}(\text{obv.}) = \mathbf{c}(\text{rev.})$.

HEXAGONAL, Laue classes $6/m$ and $6/mmm$

Reflection conditions				Laue class							
				$6/m$			$6/mmm\ (6/m\ 2/m\ 2/m)$				
				Point group							
$h\bar{h}0l$	$hh\overline{2h}l$	$000l$	Extinction symbol	6	$\bar{6}$	$6/m$	622	$6mm$	$\bar{6}2m$ $\bar{6}m2$	$6/mmm$
			$P---$	$P6$ (168)	$P\bar{6}$ (174)	$P6/m$ (175)	$P622$ (177)	$P6mm$ (183)	$P\bar{6}2m$ (189) $P\bar{6}m2$ (187)	$P6/mmm$ (191)
		l	$P6_3--$	$P6_3$ (173)		$P6_3/m$ (176)	$P6_322$ (182)			
		$l=3n$	$P6_2--$	$\left\{\begin{array}{l}P6_2\,(171)\\P6_4\,(172)\end{array}\right\}**$			$\left\{\begin{array}{l}P6_222\,(180)\\P6_422\,(181)\end{array}\right\}**$			
		$l=6n$	$P6_1--$	$\left\{\begin{array}{l}P6_1\,(169)\\P6_5\,(170)\end{array}\right\}**$			$\left\{\begin{array}{l}P6_122\,(178)\\P6_522\,(179)\end{array}\right\}**$			
	l	l	$P--c$					$P6_3mc$ (186)	$P\bar{6}2c$ (190)	$P6_3/mmc$ (194)
l		l	$P-c-$					$P6_3cm$ (185)	$P\bar{6}c2$ (188)	$P6_3/mcm$ (193)
l	l	l	$P-cc$					$P6cc$ (184)		$P6/mcc$ (192)

** Pair of enantiomorphic space groups, *cf.* Section 3.1.5.

Table 3.1.4.1. *Reflection conditions, diffraction symbols and possible space groups (cont.)*

CUBIC, Laue classes $m\bar{3}$ and $m\bar{3}m$

Reflection conditions (Indices are permutable, apart from space group No. 205) ††				Extinction symbol	Laue class				
					$m\bar{3}$ $(2/m\,\bar{3})$		$m\bar{3}m$ $(4/m\,\bar{3}\,2/m)$		
					Point group				
hkl	$0kl$	hhl	$00l$		23	$m\bar{3}$	432	$\bar{4}3m$	$m\bar{3}m$
				$P---$	$P23$ (195)	$Pm\bar{3}$ (200)	$P432$ (207)	$P\bar{4}3m$ (215)	$Pm\bar{3}m$ (221)
			l	$\begin{cases}P2_1--\\P4_2--\end{cases}$	$P2_13$ (198)		$P4_232$ (208)		
		$l=4n$		$P4_1--$			$\begin{cases}P4_132\ (213)\\P4_332\ (212)\end{cases}$‡‡		
		l	l	$P--n$				$P\bar{4}3n$ (218)	$Pm\bar{3}n$ (223)
	k††		l	$Pa--$		$Pa\bar{3}$ (205)			
	$k+l$		l	$Pn--$		$Pn\bar{3}$ (201)			$Pn\bar{3}m$ (224)
	$k+l$	l	l	$Pn-n$					$Pn\bar{3}n$ (222)
$h+k+l$	$k+l$	l	l	$I---$	$\begin{bmatrix}I23\ (197)\\I2_13\ (199)\end{bmatrix}$§§	$Im\bar{3}$ (204)	$I432$ (211)	$I\bar{4}3m$ (217)	$Im\bar{3}m$ (229)
$h+k+l$	$k+l$	l	$l=4n$	$I4_1--$			$I4_132$ (214)		
$h+k+l$	$k+l$	$2h+l=4n,l$	$l=4n$	$I--d$				$I\bar{4}3d$ (220)	
$h+k+l$	k,l	l	l	$Ia--$		$Ia\bar{3}$ (206)			
$h+k+l$	k,l	$2h+l=4n,l$	$l=4n$	$Ia-d$					$Ia\bar{3}d$ (230)
$h+k,h+l,k+l$	k,l	$h+l$	l	$F---$	$F23$ (196)	$Fm\bar{3}$ (202)	$F432$ (209)	$F\bar{4}3m$ (216)	$Fm\bar{3}m$ (225)
$h+k,h+l,k+l$	k,l	$h+l$	$l=4n$	$F4_1--$			$F4_132$ (210)		
$h+k,h+l,k+l$	k,l	h,l	l	$F--c$				$F\bar{4}3c$ (219)	$Fm\bar{3}c$ (226)
$h+k,h+l,k+l$	$k+l=4n,k,l$	$h+l$	$l=4n$	$Fd--$		$Fd\bar{3}$ (203)			$Fd\bar{3}m$ (227)
$h+k,h+l,k+l$	$k+l=4n,k,l$	h,l	$l=4n$	$Fd-c$					$Fd\bar{3}c$ (228)

†† For No. 205, only cyclic permutations are permitted. Conditions are $0kl$: $k=2n$; $h0l$: $l=2n$; $hk0$: $h=2n$.

‡‡ Pair of enantiomorphic space groups, *cf.* Section 3.1.5.

§§ Pair of space groups with common point group and symmetry elements but differing in the relative location of these elements.

3.1.6.2. *Point-group determination by methods other than the use of X-ray diffraction*

This is discussed in Chapter 10.2. In favourable cases, suitably chosen methods can prove the absence of an inversion centre or a mirror plane.

3.1.6.3. *Study of X-ray intensity distributions*

X-ray data can give a strong clue to the presence or absence of an inversion centre if not only the symmetry of the diffraction pattern but also the distribution of the intensities of the reflection spots is taken into account. Methods have been developed by Wilson and others that involve a statistical examination of certain groups of reflections. For a textbook description, see Lipson & Cochran (1966) and Wilson (1970). In this way, the presence of an inversion centre in a three-dimensional structure or in certain projections can be tested. Usually it is difficult, however, to obtain reliable conclusions from projection data. The same applies to crystals possessing pseudo-symmetry, such as a centrosymmetric arrangement of heavy atoms in a noncentrosymmetric structure. Several computer programs performing the statistical analysis of the diffraction intensities are available.

3.1.6.4. *Consideration of maxima in Patterson syntheses*

The application of Patterson syntheses for space-group determination is described by Buerger (1950, 1959).

3.1.6.5. *Anomalous dispersion*

Friedel's rule, $|F(hkl)|^2 = |F(\bar{h}\bar{k}\bar{l})|^2$, does not hold for non-centrosymmetric crystals containing atoms showing anomalous dispersion. The difference between these intensities becomes particularly strong when use is made of a wavelength near the resonance level (absorption edge) of a particular atom in the crystal. Synchrotron radiation, from which a wide variety of wavelengths can be chosen, may be used for this purpose. In such cases, the diffraction pattern reveals the symmetry of the actual point group of the crystal (including the orientation of the point group with respect to the lattice).

3.1.6.6. *Summary*

One or more of the methods discussed above may reveal whether or not the point group of the crystal has an inversion centre. With this information, in addition to the diffraction symbol, 192 space groups can be uniquely identified. The rest consist of the eleven pairs of enantiomorphic space groups, the two 'special pairs' and six further ambiguities: 3 in the orthorhombic system (Nos. 26 & 28, 35 & 38, 36 & 40), 2 in the tetragonal system (Nos. 111 & 115, 119 & 121), and 1 in the hexagonal system (Nos. 187 & 189). If not only the point group but also its orientation with respect to the lattice can be determined, the six ambiguities can be resolved. This implies that 204 space groups can be uniquely identified, the only exceptions being the eleven pairs of enantiomorphic space groups and the two 'special pairs'.

References

Buerger, M. J. (1935). *The application of plane groups to the interpretation of Weissenberg photographs. Z. Kristallogr.* **91**, 255–289.

Buerger, M. J. (1942). *X-ray crystallography*, Chap. 22. New York: Wiley.

Buerger, M. J. (1950). *The crystallographic symmetries determinable by X-ray diffraction. Proc. Natl Acad. Sci. USA*, **36**, 324–329.

Buerger, M. J. (1959). *Vector space*, pp. 167–168. New York: Wiley.

Buerger, M. J. (1960). *Crystal-structure analysis*, Chap. 5. New York: Wiley.

Buerger, M. J. (1969). *Diffraction symbols*. Chap. 3 of *Physics of the solid state*, edited by S. Balakrishna, pp. 27–42. London: Academic Press.

Catti, M. & Ferraris, G. (1976). *Twinning by merohedry and X-ray crystal structure determination. Acta Cryst.* A**32**, 163–165.

Crystal Data (1972). Vol. I, General Editors J. D. H. Donnay & H. M. Ondik, Supplement II, pp. S41–52. Washington: National Bureau of Standards.

Donnay, J. D. H. & Harker, D. (1940). *Nouvelles tables d'extinctions pour les 230 groupes de recouvrements cristallographiques. Nat. Can.* **67**, 33–69, 160.

Donnay, J. D. H. & Kennard, O. (1964). *Diffraction symbols. Acta Cryst.* **17**, 1337–1340.

Friedel, M. G. (1913). *Sur les symétries cristallines que peut révéler la diffraction des rayons Röntgen. C. R. Acad. Sci. Paris*, **157**, 1533–1536.

International Tables for X-ray Crystallography (1952; 1969). Vol. I, edited by N. F. M. Henry & K. Lonsdale. Birmingham: Kynoch Press. [Abbreviated as *IT* (1952) and *IT* (1969).]

Koch, E. (1999). *Twinning. International Tables for Crystallography* Vol. C, 2nd ed., edited by A. J. C. Wilson & E. Prince, Chap. 1.3. Dordrecht: Kluwer Academic Publishers.

Lipson, H. & Cochran, W. (1966). *The determination of crystal structures*, Chaps. 3 and 4.4. London: Bell.

Perez-Mato, J. M. & Iglesias, J. E. (1977). *On simple and double diffraction enhancement of symmetry. Acta Cryst.* A**33**, 466–474.

Wilson, A. J. C. (1970). *Elements of X-ray crystallography*, Chap. 8. Reading, MA: Addison Wesley.

5. TRANSFORMATIONS IN CRYSTALLOGRAPHY

By H. Arnold

5.1. Transformations of the coordinate system (unit-cell transformations)

H. Arnold

5.1.1. Introduction

There are two main uses of transformations in crystallography.

(i) *Transformation of the coordinate system* and the unit cell while keeping the crystal at rest. This aspect forms the main topic of the present part. Transformations of coordinate systems are useful when nonconventional descriptions of a crystal structure are considered, for instance in the study of relations between different structures, of phase transitions and of group–subgroup relations. Unit-cell transformations occur particularly frequently when different settings or cell choices of monoclinic, orthorhombic or rhombohedral space groups are to be compared or when 'reduced cells' are derived.

(ii) Description of the *symmetry operations* (motions) of an object (crystal structure). This involves the transformation of the coordinates of a point or the components of a position vector while keeping the coordinate system unchanged. Symmetry operations are treated in Chapter 8.1 and Part 11. They are briefly reviewed in Chapter 5.2.

5.1.2. Matrix notation

Throughout this volume, matrices are written in the following notation:

As (1 × 3) row matrices:

$(\mathbf{a}, \mathbf{b}, \mathbf{c})$	the basis vectors of direct space
(h, k, l)	the Miller indices of a plane (or a set of planes) in direct space or the coordinates of a point in reciprocal space

As (3 × 1) or (4 × 1) column matrices:

$x = (x/y/z)$	the coordinates of a point in direct space
$(\mathbf{a}^*/\mathbf{b}^*/\mathbf{c}^*)$	the basis vectors of reciprocal space
$(u/v/w)$	the indices of a direction in direct space
$\boldsymbol{p} = (p_1/p_2/p_3)$	the components of a shift vector from origin O to the new origin O'
$\boldsymbol{q} = (q_1/q_2/q_3)$	the components of an inverse origin shift from origin O' to origin O, with $\boldsymbol{q} = -\boldsymbol{P}^{-1}\boldsymbol{p}$
$\boldsymbol{w} = (w_1/w_2/w_3)$	the translation part of a symmetry operation W in direct space
$\mathbb{x} = (x/y/z/1)$	the augmented (4×1) column matrix of the coordinates of a point in direct space

As (3 × 3) or (4 × 4) square matrices:

$\boldsymbol{P}, \boldsymbol{Q} = \boldsymbol{P}^{-1}$	linear parts of an affine transformation; if $\boldsymbol{P}$ is applied to a (1×3) row matrix, $\boldsymbol{Q}$ must be applied to a (3×1) column matrix, and *vice versa*
$\boldsymbol{W}$	the rotation part of a symmetry operation W in direct space
$\mathbb{P} = \begin{pmatrix} \boldsymbol{P} & \boldsymbol{p} \\ \boldsymbol{o} & 1 \end{pmatrix}$	the augmented affine (4×4) transformation matrix, with $\boldsymbol{o} = (0,0,0)$
$\mathbb{Q} = \begin{pmatrix} \boldsymbol{Q} & \boldsymbol{q} \\ \boldsymbol{o} & 1 \end{pmatrix}$	the augmented affine (4×4) transformation matrix, with $\mathbb{Q} = \mathbb{P}^{-1}$
$\mathbb{W} = \begin{pmatrix} \boldsymbol{W} & \boldsymbol{w} \\ \boldsymbol{o} & 1 \end{pmatrix}$	the augmented (4×4) matrix of a symmetry operation in direct space (*cf.* Chapter 8.1 and Part 11).

5.1.3. General transformation

Here the crystal structure is considered to be at rest, whereas the coordinate system and the unit cell are changed. Specifically, a point X in a crystal is defined with respect to the basis vectors $\mathbf{a}, \mathbf{b}, \mathbf{c}$ and the origin O by the coordinates x, y, z, *i.e.* the position vector $\mathbf{r}$ of point X is given by

$$\mathbf{r} = x\mathbf{a} + y\mathbf{b} + z\mathbf{c}$$
$$= (\mathbf{a}, \mathbf{b}, \mathbf{c}) \begin{pmatrix} x \\ y \\ z \end{pmatrix}.$$

The same point X is given with respect to a new coordinate system, *i.e.* the new basis vectors $\mathbf{a}', \mathbf{b}', \mathbf{c}'$ and the new origin O' (Fig. 5.1.3.1), by the position vector

$$\mathbf{r}' = x'\mathbf{a}' + y'\mathbf{b}' + z'\mathbf{c}'.$$

In this section, the relations between the primed and unprimed quantities are treated.

The general transformation (affine transformation) of the coordinate system consists of two parts, a linear part and a shift of origin. The (3×3) matrix $\boldsymbol{P}$ of the linear part and the (3×1) column matrix $\boldsymbol{p}$, containing the components of the shift vector $\boldsymbol{p}$, define the transformation uniquely. It is represented by the symbol $(\boldsymbol{P}, \boldsymbol{p})$.

(i) The *linear part* implies a change of orientation or length or both of the basis vectors $\mathbf{a}, \mathbf{b}, \mathbf{c}$, *i.e.*

$$(\mathbf{a}', \mathbf{b}', \mathbf{c}') = (\mathbf{a}, \mathbf{b}, \mathbf{c})\boldsymbol{P}$$
$$= (\mathbf{a}, \mathbf{b}, \mathbf{c}) \begin{pmatrix} P_{11} & P_{12} & P_{13} \\ P_{21} & P_{22} & P_{23} \\ P_{31} & P_{32} & P_{33} \end{pmatrix}$$
$$= (P_{11}\mathbf{a} + P_{21}\mathbf{b} + P_{31}\mathbf{c},$$
$$P_{12}\mathbf{a} + P_{22}\mathbf{b} + P_{32}\mathbf{c},$$
$$P_{13}\mathbf{a} + P_{23}\mathbf{b} + P_{33}\mathbf{c}).$$

For a pure linear transformation, the shift vector $\mathbf{p}$ is zero and the symbol is $(\boldsymbol{P}, \boldsymbol{o})$.

The determinant of $\boldsymbol{P}$, $\det(\boldsymbol{P})$, should be positive. If $\det(\boldsymbol{P})$ is negative, a right-handed coordinate system is transformed into a left-handed one (or *vice versa*). If $\det(\boldsymbol{P}) = 0$, the new basis vectors are linearly dependent and do not form a complete coordinate system.

In this chapter, transformations in three-dimensional space are treated. A change of the basis vectors in two dimensions, *i.e.* of the basis vectors $\mathbf{a}$ and $\mathbf{b}$, can be considered as a three-dimensional transformation with invariant $\mathbf{c}$ axis. This is achieved by setting $P_{33} = 1$ and $P_{13} = P_{23} = P_{31} = P_{32} = 0$.

(ii) A *shift of origin* is defined by the shift vector

$$\mathbf{p} = p_1\mathbf{a} + p_2\mathbf{b} + p_3\mathbf{c}.$$

The basis vectors $\mathbf{a}, \mathbf{b}, \mathbf{c}$ are fixed at the origin O; the new basis vectors are fixed at the new origin O' which has the coordinates p_1, p_2, p_3 in the old coordinate system (Fig. 5.1.3.1).

For a pure origin shift, the basis vectors do not change their lengths or orientations. In this case, the transformation matrix $\boldsymbol{P}$ is the unit matrix $\boldsymbol{I}$ and the symbol of the pure shift becomes $(\boldsymbol{I}, \boldsymbol{p})$.

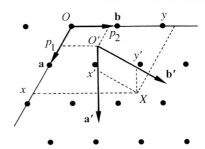

Fig. 5.1.3.1. General affine transformation, consisting of a shift of origin from O to O' by a shift vector $\mathbf{p}$ with components p_1 and p_2 and a change of basis from $\mathbf{a}, \mathbf{b}$ to $\mathbf{a}', \mathbf{b}'$. This implies a change in the coordinates of the point X from x, y to x', y'.

Also, the inverse matrices of P and p are needed. They are

$$Q = P^{-1}$$

and

$$q = -P^{-1}p.$$

The matrix q consists of the components of the negative shift vector $\mathbf{q}$ which refer to the coordinate system $\mathbf{a}', \mathbf{b}', \mathbf{c}'$, i.e.

$$\mathbf{q} = q_1\mathbf{a}' + q_2\mathbf{b}' + q_3\mathbf{c}'.$$

Thus, the transformation (Q, q) is the inverse transformation of (P, p). Applying (Q, q) to the basis vectors $\mathbf{a}', \mathbf{b}', \mathbf{c}'$ and the origin O', the old basis vectors $\mathbf{a}, \mathbf{b}, \mathbf{c}$ with origin O are obtained.

For a two-dimensional transformation of $\mathbf{a}'$ and $\mathbf{b}'$, some elements of Q are set as follows: $Q_{33} = 1$ and $Q_{13} = Q_{23} = Q_{31} = Q_{32} = 0$.

The quantities which transform in the same way as the basis vectors $\mathbf{a}, \mathbf{b}, \mathbf{c}$ are called *covariant* quantities and are written as row matrices. They are:

the *Miller indices of a plane* (or a set of planes), (hkl), in direct space and

the *coordinates of a point in reciprocal space*, h, k, l.

Both are transformed by

$$(h', k', l') = (h, k, l)P.$$

Usually, the Miller indices are made relative prime before and after the transformation.

The quantities which are covariant with respect to the basis vectors $\mathbf{a}, \mathbf{b}, \mathbf{c}$ are *contravariant* with respect to the basis vectors $\mathbf{a}^*, \mathbf{b}^*, \mathbf{c}^*$ of reciprocal space.

The *basis vectors of reciprocal space* are written as a column matrix and their transformation is achieved by the matrix Q:

$$
\begin{pmatrix} \mathbf{a}^{*\prime} \\ \mathbf{b}^{*\prime} \\ \mathbf{c}^{*\prime} \end{pmatrix} = Q \begin{pmatrix} \mathbf{a}^* \\ \mathbf{b}^* \\ \mathbf{c}^* \end{pmatrix}
$$

$$
= \begin{pmatrix} Q_{11} & Q_{12} & Q_{13} \\ Q_{21} & Q_{22} & Q_{23} \\ Q_{31} & Q_{32} & Q_{33} \end{pmatrix} \begin{pmatrix} \mathbf{a}^* \\ \mathbf{b}^* \\ \mathbf{c}^* \end{pmatrix}
$$

$$
= \begin{pmatrix} Q_{11}\mathbf{a}^* + Q_{12}\mathbf{b}^* + Q_{13}\mathbf{c}^* \\ Q_{21}\mathbf{a}^* + Q_{22}\mathbf{b}^* + Q_{23}\mathbf{c}^* \\ Q_{31}\mathbf{a}^* + Q_{32}\mathbf{b}^* + Q_{33}\mathbf{c}^* \end{pmatrix}.
$$

The inverse transformation is obtained by the inverse matrix

$$P = Q^{-1}:$$

$$
\begin{pmatrix} \mathbf{a}^* \\ \mathbf{b}^* \\ \mathbf{c}^* \end{pmatrix} = P \begin{pmatrix} \mathbf{a}^{*\prime} \\ \mathbf{b}^{*\prime} \\ \mathbf{c}^{*\prime} \end{pmatrix}.
$$

These transformation rules apply also to the quantities covariant with respect to the basis vectors $\mathbf{a}^*, \mathbf{b}^*, \mathbf{c}^*$ and contravariant with respect to $\mathbf{a}, \mathbf{b}, \mathbf{c}$, which are written as column matrices. They are the *indices of a direction* in direct space, $[uvw]$, which are transformed by

$$
\begin{pmatrix} u' \\ v' \\ w' \end{pmatrix} = Q \begin{pmatrix} u \\ v \\ w \end{pmatrix}.
$$

In contrast to all quantities mentioned above, the *components of a position vector* $\mathbf{r}$ or the *coordinates of a point X in direct space* x, y, z depend also on the shift of the origin in direct space. The general (affine) transformation is given by

$$
\begin{pmatrix} x' \\ y' \\ z' \end{pmatrix} = Q \begin{pmatrix} x \\ y \\ z \end{pmatrix} + q
$$

$$
= \begin{pmatrix} Q_{11}x + Q_{12}y + Q_{13}z + q_1 \\ Q_{21}x + Q_{22}y + Q_{23}z + q_2 \\ Q_{31}x + Q_{32}y + Q_{33}z + q_3 \end{pmatrix}.
$$

Example

If no shift of origin is applied, *i.e.* $\mathbf{p} = \mathbf{q} = \mathbf{o}$, the position vector $\mathbf{r}$ of point X is transformed by

$$
\mathbf{r}' = (\mathbf{a}, \mathbf{b}, \mathbf{c})PQ \begin{pmatrix} x \\ y \\ z \end{pmatrix} = (\mathbf{a}', \mathbf{b}', \mathbf{c}') \begin{pmatrix} x' \\ y' \\ z' \end{pmatrix}.
$$

In this case, $\mathbf{r} = \mathbf{r}'$, *i.e.* the position vector is invariant, although the basis vectors and the components are transformed. For a pure shift of origin, *i.e.* $P = Q = I$, the transformed position vector $\mathbf{r}'$ becomes

$$
\begin{aligned}
\mathbf{r}' &= (x + q_1)\mathbf{a} + (y + q_2)\mathbf{b} + (z + q_3)\mathbf{c} \\
&= \mathbf{r} + q_1\mathbf{a} + q_2\mathbf{b} + q_3\mathbf{c} \\
&= (x - p_1)\mathbf{a} + (y - p_2)\mathbf{b} + (z - p_3)\mathbf{c} \\
&= \mathbf{r} - p_1\mathbf{a} - p_2\mathbf{b} - p_3\mathbf{c}.
\end{aligned}
$$

Here the transformed vector $\mathbf{r}'$ is no longer identical with $\mathbf{r}$.

It is convenient to introduce the augmented (4×4) matrix $\mathbb{Q}$ which is composed of the matrices Q and q in the following manner (*cf.* Chapter 8.1):

$$
\mathbb{Q} = \begin{pmatrix} Q & q \\ o & 1 \end{pmatrix} = \begin{pmatrix} Q_{11} & Q_{12} & Q_{13} & q_1 \\ Q_{21} & Q_{22} & Q_{23} & q_2 \\ Q_{31} & Q_{32} & Q_{33} & q_3 \\ 0 & 0 & 0 & 1 \end{pmatrix}
$$

with o the (1×3) row matrix containing zeros. In this notation, the transformed coordinates x', y', z' are obtained by

Table 5.1.3.1. *Selected 3 × 3 transformation matrices P and $Q = P^{-1}$*

For inverse transformations (against the arrow) replace P by Q and *vice versa.*

Transformation	P	$Q = P^{-1}$	Crystal system
$\mathbf{c} \to \frac{1}{2}\mathbf{c}$	$\begin{pmatrix} 1 & 0 & 0 \\ 0 & 1 & 0 \\ 0 & 0 & \frac{1}{2} \end{pmatrix}$	$\begin{pmatrix} 1 & 0 & 0 \\ 0 & 1 & 0 \\ 0 & 0 & 2 \end{pmatrix}$	All systems
$\mathbf{b} \to \frac{1}{2}\mathbf{b}$	$\begin{pmatrix} 1 & 0 & 0 \\ 0 & \frac{1}{2} & 0 \\ 0 & 0 & 1 \end{pmatrix}$	$\begin{pmatrix} 1 & 0 & 0 \\ 0 & 2 & 0 \\ 0 & 0 & 1 \end{pmatrix}$	All systems
$\mathbf{a} \to \frac{1}{2}\mathbf{a}$	$\begin{pmatrix} \frac{1}{2} & 0 & 0 \\ 0 & 1 & 0 \\ 0 & 0 & 1 \end{pmatrix}$	$\begin{pmatrix} 2 & 0 & 0 \\ 0 & 1 & 0 \\ 0 & 0 & 1 \end{pmatrix}$	All systems
Cell choice 1 → cell choice 2: $\begin{cases} P \to P \\ C \to A \end{cases}$ Cell choice 2 → cell choice 3: $\begin{cases} P \to P \\ A \to I \end{cases}$ Unique axis $\mathbf{b}$ invariant Cell choice 3 → cell choice 1: $\begin{cases} P \to P \\ I \to C \end{cases}$ (Fig. 5.1.3.2a)	$\begin{pmatrix} \bar{1} & 0 & 1 \\ 0 & 1 & 0 \\ \bar{1} & 0 & 0 \end{pmatrix}$	$\begin{pmatrix} 0 & 0 & \bar{1} \\ 0 & 1 & 0 \\ 1 & 0 & \bar{1} \end{pmatrix}$	Monoclinic (*cf.* Section 2.2.16)
Cell choice 1 → cell choice 2: $\begin{cases} P \to P \\ A \to B \end{cases}$ Cell choice 2 → cell choice 3: $\begin{cases} P \to P \\ B \to I \end{cases}$ Unique axis $\mathbf{c}$ invariant Cell choice 3 → cell choice 1: $\begin{cases} P \to P \\ I \to A \end{cases}$ (Fig. 5.1.3.2b)	$\begin{pmatrix} 0 & \bar{1} & 0 \\ 1 & \bar{1} & 0 \\ 0 & 0 & 1 \end{pmatrix}$	$\begin{pmatrix} \bar{1} & 1 & 0 \\ \bar{1} & 0 & 0 \\ 0 & 0 & 1 \end{pmatrix}$	Monoclinic (*cf.* Section 2.2.16)
Cell choice 1 → cell choice 2: $\begin{cases} P \to P \\ B \to C \end{cases}$ Cell choice 2 → cell choice 3: $\begin{cases} P \to P \\ C \to I \end{cases}$ Unique axis $\mathbf{a}$ invariant Cell choice 3 → cell choice 1: $\begin{cases} P \to P \\ I \to B \end{cases}$ (Fig. 5.1.3.2c)	$\begin{pmatrix} 1 & 0 & 0 \\ 0 & 0 & \bar{1} \\ 0 & 1 & \bar{1} \end{pmatrix}$	$\begin{pmatrix} 1 & 0 & 0 \\ 0 & \bar{1} & 1 \\ 0 & \bar{1} & 0 \end{pmatrix}$	Monoclinic (*cf.* Section 2.2.16)
Unique axis $\mathbf{b}$ → unique axis $\mathbf{c}$ Cell choice 1: $\begin{cases} P \to P \\ C \to A \end{cases}$ Cell choice 2: $\begin{cases} P \to P \\ A \to B \end{cases}$ Cell choice invariant Cell choice 3: $\begin{cases} P \to P \\ I \to I \end{cases}$	$\begin{pmatrix} 0 & 1 & 0 \\ 0 & 0 & 1 \\ 1 & 0 & 0 \end{pmatrix}$	$\begin{pmatrix} 0 & 0 & 1 \\ 1 & 0 & 0 \\ 0 & 1 & 0 \end{pmatrix}$	Monoclinic (*cf.* Section 2.2.16)
Unique axis $\mathbf{b}$ → unique axis $\mathbf{a}$ Cell choice 1: $\begin{cases} P \to P \\ C \to B \end{cases}$ Cell choice 2: $\begin{cases} P \to P \\ A \to C \end{cases}$ Cell choice invariant Cell choice 3: $\begin{cases} P \to P \\ I \to I \end{cases}$	$\begin{pmatrix} 0 & 0 & 1 \\ 1 & 0 & 0 \\ 0 & 1 & 0 \end{pmatrix}$	$\begin{pmatrix} 0 & 1 & 0 \\ 0 & 0 & 1 \\ 1 & 0 & 0 \end{pmatrix}$	Monoclinic (*cf.* Section 2.2.16)
Unique axis $\mathbf{c}$ → unique axis $\mathbf{a}$ Cell choice 1: $\begin{cases} P \to P \\ A \to B \end{cases}$ Cell choice 2: $\begin{cases} P \to P \\ B \to C \end{cases}$ Cell choice invariant Cell choice 3: $\begin{cases} P \to P \\ I \to I \end{cases}$	$\begin{pmatrix} 0 & 1 & 0 \\ 0 & 0 & 1 \\ 1 & 0 & 0 \end{pmatrix}$	$\begin{pmatrix} 0 & 0 & 1 \\ 1 & 0 & 0 \\ 0 & 1 & 0 \end{pmatrix}$	Monoclinic (*cf.* Section 2.2.16)
$I \to P$ (Fig. 5.1.3.3) $\quad \frac{1}{2}(\mathbf{a}+\mathbf{b}+\mathbf{c}) \to (\mathbf{a}'+\mathbf{b}'+\mathbf{c}')$	$\begin{pmatrix} \bar{1} & \frac{1}{2} & \frac{1}{2} \\ \frac{1}{2} & \bar{1} & \frac{1}{2} \\ \frac{1}{2} & \frac{1}{2} & \bar{1} \end{pmatrix}$	$\begin{pmatrix} 0 & 1 & 1 \\ 1 & 0 & 1 \\ 1 & 1 & 0 \end{pmatrix}$	Orthorhombic Tetragonal Cubic

Table 5.1.3.1. *Selected 3×3 transformation matrices P and $Q = P^{-1}$ (cont.)*

Transformation	P	$Q = P^{-1}$	Crystal system
$F \to P$ (Fig. 5.1.3.4) $(\mathbf{a}+\mathbf{b}+\mathbf{c})$ invariant vector	$\begin{pmatrix} 0 & \frac{1}{2} & \frac{1}{2} \\ \frac{1}{2} & 0 & \frac{1}{2} \\ \frac{1}{2} & \frac{1}{2} & 0 \end{pmatrix}$	$\begin{pmatrix} \bar{1} & 1 & 1 \\ 1 & \bar{1} & 1 \\ 1 & 1 & \bar{1} \end{pmatrix}$	Orthorhombic Tetragonal Cubic
$(\mathbf{b},\mathbf{a},\bar{\mathbf{c}}) \to (\mathbf{a},\mathbf{b},\mathbf{c})$	$\begin{pmatrix} 0 & 1 & 0 \\ 1 & 0 & 0 \\ 0 & 0 & \bar{1} \end{pmatrix}$	$\begin{pmatrix} 0 & 1 & 0 \\ 1 & 0 & 0 \\ 0 & 0 & \bar{1} \end{pmatrix}$	Unconventional orthorhombic setting
$(\mathbf{c},\mathbf{a},\mathbf{b}) \to (\mathbf{a},\mathbf{b},\mathbf{c})$	$\begin{pmatrix} 0 & 0 & 1 \\ 1 & 0 & 0 \\ 0 & 1 & 0 \end{pmatrix}$	$\begin{pmatrix} 0 & 1 & 0 \\ 0 & 0 & 1 \\ 1 & 0 & 0 \end{pmatrix}$	Unconventional orthorhombic setting
$(\bar{\mathbf{c}},\mathbf{b},\mathbf{a}) \to (\mathbf{a},\mathbf{b},\mathbf{c})$	$\begin{pmatrix} 0 & 0 & \bar{1} \\ 0 & 1 & 0 \\ 1 & 0 & 0 \end{pmatrix}$	$\begin{pmatrix} 0 & 0 & 1 \\ 0 & 1 & 0 \\ \bar{1} & 0 & 0 \end{pmatrix}$	Unconventional orthorhombic setting
$(\mathbf{b},\mathbf{c},\mathbf{a}) \to (\mathbf{a},\mathbf{b},\mathbf{c})$	$\begin{pmatrix} 0 & 1 & 0 \\ 0 & 0 & 1 \\ 1 & 0 & 0 \end{pmatrix}$	$\begin{pmatrix} 0 & 0 & 1 \\ 1 & 0 & 0 \\ 0 & 1 & 0 \end{pmatrix}$	Unconventional orthorhombic setting
$(\mathbf{a},\bar{\mathbf{c}},\mathbf{b}) \to (\mathbf{a},\mathbf{b},\mathbf{c})$	$\begin{pmatrix} 1 & 0 & 0 \\ 0 & 0 & \bar{1} \\ 0 & 1 & 0 \end{pmatrix}$	$\begin{pmatrix} 1 & 0 & 0 \\ 0 & 0 & 1 \\ 0 & \bar{1} & 0 \end{pmatrix}$	Unconventional orthorhombic setting
$\left.\begin{array}{l} P \to C_1 \\ I \to F_1 \end{array}\right\}$ (Fig. 5.1.3.5) $\mathbf{c}$ axis invariant	$\begin{pmatrix} 1 & 1 & 0 \\ \bar{1} & 1 & 0 \\ 0 & 0 & 1 \end{pmatrix}$	$\begin{pmatrix} \frac{1}{2} & \bar{\frac{1}{2}} & 0 \\ \frac{1}{2} & \frac{1}{2} & 0 \\ 0 & 0 & 1 \end{pmatrix}$	Tetragonal (*cf.* Section 4.3.4)
$\left.\begin{array}{l} P \to C_2 \\ I \to F_2 \end{array}\right\}$ (Fig. 5.1.3.5) $\mathbf{c}$ axis invariant	$\begin{pmatrix} 1 & \bar{1} & 0 \\ 1 & 1 & 0 \\ 0 & 0 & 1 \end{pmatrix}$	$\begin{pmatrix} \frac{1}{2} & \frac{1}{2} & 0 \\ \bar{\frac{1}{2}} & \frac{1}{2} & 0 \\ 0 & 0 & 1 \end{pmatrix}$	Tetragonal (*cf.* Section 4.3.4)
Primitive rhombohedral cell $\to$ triple hexagonal cell R_1, obverse setting (Fig. 5.1.3.6*c*)	$\begin{pmatrix} 1 & 0 & 1 \\ \bar{1} & 1 & 1 \\ 0 & \bar{1} & 1 \end{pmatrix}$	$\begin{pmatrix} \frac{2}{3} & \bar{\frac{1}{3}} & \bar{\frac{1}{3}} \\ \frac{1}{3} & \frac{1}{3} & \bar{\frac{2}{3}} \\ \frac{1}{3} & \frac{1}{3} & \frac{1}{3} \end{pmatrix}$	Rhombohedral space groups (*cf.* Section 4.3.5)
Primitive rhombohedral cell $\to$ triple hexagonal cell R_2, obverse setting (Fig. 5.1.3.6*c*)	$\begin{pmatrix} 0 & \bar{1} & 1 \\ 1 & 0 & 1 \\ \bar{1} & 1 & 1 \end{pmatrix}$	$\begin{pmatrix} \bar{\frac{1}{3}} & \frac{2}{3} & \bar{\frac{1}{3}} \\ \bar{\frac{2}{3}} & \frac{1}{3} & \frac{1}{3} \\ \frac{1}{3} & \frac{1}{3} & \frac{1}{3} \end{pmatrix}$	Rhombohedral space groups (*cf.* Section 4.3.5)
Primitive rhombohedral cell $\to$ triple hexagonal cell R_3, obverse setting (Fig. 5.1.3.6*c*)	$\begin{pmatrix} \bar{1} & 1 & 1 \\ 0 & \bar{1} & 1 \\ 1 & 0 & 1 \end{pmatrix}$	$\begin{pmatrix} \bar{\frac{1}{3}} & \bar{\frac{1}{3}} & \frac{2}{3} \\ \frac{1}{3} & \bar{\frac{2}{3}} & \frac{1}{3} \\ \frac{1}{3} & \frac{1}{3} & \frac{1}{3} \end{pmatrix}$	Rhombohedral space groups (*cf.* Section 4.3.5)
Primitive rhombohedral cell $\to$ triple hexagonal cell R_1, reverse setting (Fig. 5.1.3.6*d*)	$\begin{pmatrix} \bar{1} & 0 & 1 \\ 1 & \bar{1} & 1 \\ 0 & 1 & 1 \end{pmatrix}$	$\begin{pmatrix} \bar{\frac{2}{3}} & \frac{1}{3} & \frac{1}{3} \\ \bar{\frac{1}{3}} & \bar{\frac{1}{3}} & \frac{2}{3} \\ \frac{1}{3} & \frac{1}{3} & \frac{1}{3} \end{pmatrix}$	Rhombohedral space groups (*cf.* Section 4.3.5)
Primitive rhombohedral cell $\to$ triple hexagonal cell R_2, reverse setting (Fig. 5.1.3.6*d*)	$\begin{pmatrix} 0 & 1 & 1 \\ \bar{1} & 0 & 1 \\ 1 & \bar{1} & 1 \end{pmatrix}$	$\begin{pmatrix} \frac{1}{3} & \bar{\frac{2}{3}} & \frac{1}{3} \\ \frac{2}{3} & \bar{\frac{1}{3}} & \frac{1}{3} \\ \frac{1}{3} & \frac{1}{3} & \frac{1}{3} \end{pmatrix}$	Rhombohedral space groups (*cf.* Section 4.3.5)
Primitive rhombohedral cell $\to$ triple hexagonal cell R_3, reverse setting (Fig. 5.1.3.6*d*)	$\begin{pmatrix} 1 & \bar{1} & 1 \\ 0 & 1 & 1 \\ \bar{1} & 0 & 1 \end{pmatrix}$	$\begin{pmatrix} \frac{1}{3} & \frac{1}{3} & \bar{\frac{2}{3}} \\ \bar{\frac{1}{3}} & \frac{2}{3} & \bar{\frac{1}{3}} \\ \frac{1}{3} & \frac{1}{3} & \frac{1}{3} \end{pmatrix}$	Rhombohedral space groups (*cf.* Section 4.3.5)
Hexagonal cell $P \to$ orthohexagonal centred cell C_1 (Fig. 5.1.3.7)	$\begin{pmatrix} 1 & 1 & 0 \\ 0 & 2 & 0 \\ 0 & 0 & 1 \end{pmatrix}$	$\begin{pmatrix} 1 & \bar{\frac{1}{2}} & 0 \\ 0 & \frac{1}{2} & 0 \\ 0 & 0 & 1 \end{pmatrix}$	Trigonal Hexagonal (*cf.* Section 4.3.5)
Hexagonal cell $P \to$ orthohexagonal centred cell C_2 (Fig. 5.1.3.7)	$\begin{pmatrix} 1 & \bar{1} & 0 \\ 1 & 1 & 0 \\ 0 & 0 & 1 \end{pmatrix}$	$\begin{pmatrix} \frac{1}{2} & \frac{1}{2} & 0 \\ \bar{\frac{1}{2}} & \frac{1}{2} & 0 \\ 0 & 0 & 1 \end{pmatrix}$	Trigonal Hexagonal (*cf.* Section 4.3.5)
Hexagonal cell $P \to$ orthohexagonal centred cell C_3 (Fig. 5.1.3.7)	$\begin{pmatrix} 0 & \bar{2} & 0 \\ 1 & \bar{1} & 0 \\ 0 & 0 & 1 \end{pmatrix}$	$\begin{pmatrix} \bar{\frac{1}{2}} & 1 & 0 \\ \bar{\frac{1}{2}} & 0 & 0 \\ 0 & 0 & 1 \end{pmatrix}$	Trigonal Hexagonal (*cf.* Section 4.3.5)

Table 5.1.3.1. *Selected 3 × 3 transformation matrices* P *and* $Q = P^{-1}$ *(cont.)*

Transformation	P	$Q = P^{-1}$	Crystal system
Hexagonal cell $P \rightarrow$ triple hexagonal cell H_1 (Fig. 5.1.3.8)	$\begin{pmatrix} 1 & 1 & 0 \\ \bar{1} & 2 & 0 \\ 0 & 0 & 1 \end{pmatrix}$	$\begin{pmatrix} \frac{2}{3} & \frac{\bar{1}}{3} & 0 \\ \frac{1}{3} & \frac{1}{3} & 0 \\ 0 & 0 & 1 \end{pmatrix}$	Trigonal Hexagonal (*cf.* Section 4.3.5)
Hexagonal cell $P \rightarrow$ triple hexagonal cell H_2 (Fig. 5.1.3.8)	$\begin{pmatrix} 2 & \bar{1} & 0 \\ 1 & 1 & 0 \\ 0 & 0 & 1 \end{pmatrix}$	$\begin{pmatrix} \frac{1}{3} & \frac{1}{3} & 0 \\ \frac{\bar{1}}{3} & \frac{2}{3} & 0 \\ 0 & 0 & 1 \end{pmatrix}$	Trigonal Hexagonal (*cf.* Section 4.3.5)
Hexagonal cell $P \rightarrow$ triple hexagonal cell H_3 (Fig. 5.1.3.8)	$\begin{pmatrix} 1 & \bar{2} & 0 \\ 2 & \bar{1} & 0 \\ 0 & 0 & 1 \end{pmatrix}$	$\begin{pmatrix} \frac{\bar{1}}{3} & \frac{2}{3} & 0 \\ \frac{\bar{2}}{3} & \frac{1}{3} & 0 \\ 0 & 0 & 1 \end{pmatrix}$	Trigonal Hexagonal (*cf.* Section 4.3.5)
Hexagonal cell $P \rightarrow$ triple rhombohedral cell D_1	$\begin{pmatrix} 1 & 0 & \bar{1} \\ 0 & 1 & \bar{1} \\ 1 & 1 & 1 \end{pmatrix}$	$\begin{pmatrix} \frac{2}{3} & \frac{\bar{1}}{3} & \frac{1}{3} \\ \frac{\bar{1}}{3} & \frac{2}{3} & \frac{1}{3} \\ \frac{\bar{1}}{3} & \frac{\bar{1}}{3} & \frac{1}{3} \end{pmatrix}$	Trigonal Hexagonal (*cf.* Section 4.3.5)
Hexagonal cell $P \rightarrow$ triple rhombohedral cell D_2	$\begin{pmatrix} \bar{1} & 0 & 1 \\ 0 & \bar{1} & 1 \\ 1 & 1 & 1 \end{pmatrix}$	$\begin{pmatrix} \frac{\bar{2}}{3} & \frac{1}{3} & \frac{1}{3} \\ \frac{1}{3} & \frac{\bar{2}}{3} & \frac{1}{3} \\ \frac{1}{3} & \frac{1}{3} & \frac{1}{3} \end{pmatrix}$	Trigonal Hexagonal (*cf.* Section 4.3.5)
Triple hexagonal cell R, obverse setting $\rightarrow$ C-centred monoclinic cell, unique axis **b**, cell choice 1 (Fig. 5.1.3.9a) **c** and **b** axes invariant	$\begin{pmatrix} \frac{2}{3} & 0 & 0 \\ \frac{1}{3} & 1 & 0 \\ \frac{\bar{2}}{3} & 0 & 1 \end{pmatrix}$	$\begin{pmatrix} \frac{3}{2} & 0 & 0 \\ \frac{\bar{1}}{2} & 1 & 0 \\ 1 & 0 & 1 \end{pmatrix}$	Rhombohedral space groups (*cf.* Section 4.3.5)
Triple hexagonal cell R, obverse setting $\rightarrow$ C-centred monoclinic cell, unique axis **b**, cell choice 2 (Fig. 5.1.3.9a) **c** axis invariant	$\begin{pmatrix} \frac{\bar{1}}{3} & \bar{1} & 0 \\ \frac{\bar{1}}{3} & \bar{1} & 0 \\ \frac{\bar{2}}{3} & 0 & 1 \end{pmatrix}$	$\begin{pmatrix} \frac{\bar{3}}{2} & \frac{3}{2} & 0 \\ \frac{\bar{1}}{2} & \frac{\bar{1}}{2} & 0 \\ \bar{1} & 1 & 1 \end{pmatrix}$	Rhombohedral space groups (*cf.* Section 4.3.5)
Triple hexagonal cell R, obverse setting $\rightarrow$ C-centred monoclinic cell, unique axis **b**, cell choice 3 (Fig. 5.1.3.9a) $\mathbf{a}_h \rightarrow \mathbf{b}_m$, **c** axis invariant	$\begin{pmatrix} \frac{\bar{1}}{3} & 1 & 0 \\ \frac{\bar{2}}{3} & 0 & 0 \\ \frac{\bar{2}}{3} & 0 & 1 \end{pmatrix}$	$\begin{pmatrix} 0 & \frac{\bar{3}}{2} & 0 \\ 1 & \frac{\bar{1}}{2} & 0 \\ 0 & \bar{1} & 1 \end{pmatrix}$	Rhombohedral space groups (*cf.* Section 4.3.5)
Triple hexagonal cell R, obverse setting $\rightarrow$ A-centred monoclinic cell, unique axis **c**, cell choice 1 (Fig. 5.1.3.9b) $\mathbf{b}_h \rightarrow \mathbf{c}_m, \mathbf{c}_h \rightarrow \mathbf{a}_m$	$\begin{pmatrix} 0 & \frac{2}{3} & 0 \\ 0 & \frac{1}{3} & 1 \\ 1 & \frac{2}{3} & 0 \end{pmatrix}$	$\begin{pmatrix} 1 & 0 & 1 \\ \frac{3}{2} & 0 & 0 \\ \frac{\bar{1}}{2} & 1 & 0 \end{pmatrix}$	Rhombohedral space groups (*cf.* Section 4.3.5)
Triple hexagonal cell R, obverse setting $\rightarrow$ A-centred monoclinic cell, unique axis **c**, cell choice 2 (Fig. 5.1.3.9b) $\mathbf{c}_h \rightarrow \mathbf{a}_m$	$\begin{pmatrix} 0 & \frac{\bar{1}}{3} & \bar{1} \\ 0 & \frac{1}{3} & \bar{1} \\ 1 & \frac{2}{3} & 0 \end{pmatrix}$	$\begin{pmatrix} \bar{1} & 1 & 1 \\ \frac{3}{2} & \frac{3}{2} & 0 \\ \frac{\bar{1}}{2} & \frac{\bar{1}}{2} & 0 \end{pmatrix}$	Rhombohedral space groups (*cf.* Section 4.3.5)
Triple hexagonal cell R, obverse setting $\rightarrow$ A-centred monoclinic cell, unique axis **c**, cell choice 3 (Fig. 5.1.3.9b) $\mathbf{a}_h \rightarrow \mathbf{c}_m, \mathbf{c}_h \rightarrow \mathbf{a}_m$	$\begin{pmatrix} 0 & \frac{1}{3} & 1 \\ 0 & \frac{\bar{2}}{3} & 0 \\ 1 & \frac{2}{3} & 0 \end{pmatrix}$	$\begin{pmatrix} 0 & \bar{1} & 1 \\ 0 & \frac{\bar{3}}{2} & 0 \\ 1 & \frac{\bar{1}}{2} & 0 \end{pmatrix}$	Rhombohedral space groups (*cf.* Section 4.3.5)
Primitive rhombohedral cell $\rightarrow$ C-centred monoclinic cell, unique axis **b**, cell choice 1 (Fig. 5.1.3.10a) $[111]_r \rightarrow \mathbf{c}_m$	$\begin{pmatrix} 0 & 0 & 1 \\ \bar{1} & 1 & 1 \\ \bar{1} & \bar{1} & 1 \end{pmatrix}$	$\begin{pmatrix} 1 & \frac{\bar{1}}{2} & \frac{\bar{1}}{2} \\ 0 & \frac{1}{2} & \frac{\bar{1}}{2} \\ 1 & 0 & 0 \end{pmatrix}$	Rhombohedral space groups (*cf.* Section 4.3.5)
Primitive rhombohedral cell $\rightarrow$ C-centred monoclinic cell, unique axis **b**, cell choice 2 (Fig. 5.1.3.10a) $[111]_r \rightarrow \mathbf{c}_m$	$\begin{pmatrix} \bar{1} & \bar{1} & 1 \\ 0 & 0 & 1 \\ \bar{1} & 1 & 1 \end{pmatrix}$	$\begin{pmatrix} \frac{\bar{1}}{2} & 1 & \frac{\bar{1}}{2} \\ \frac{\bar{1}}{2} & 0 & \frac{1}{2} \\ 0 & 1 & 0 \end{pmatrix}$	Rhombohedral space groups (*cf.* Section 4.3.5)
Primitive rhombohedral cell $\rightarrow$ C-centred monoclinic cell, unique axis **b**, cell choice 3 (Fig. 5.1.3.10a) $[111]_r \rightarrow \mathbf{c}_m$	$\begin{pmatrix} \bar{1} & 1 & 1 \\ \bar{1} & \bar{1} & 1 \\ 0 & 0 & 1 \end{pmatrix}$	$\begin{pmatrix} \frac{\bar{1}}{2} & \frac{\bar{1}}{2} & 1 \\ \frac{1}{2} & \frac{\bar{1}}{2} & 0 \\ 0 & 0 & 1 \end{pmatrix}$	Rhombohedral space groups (*cf.* Section 4.3.5)
Primitive rhombohedral cell $\rightarrow$ A-centred monoclinic cell, unique axis **c**, cell choice 1 (Fig. 5.1.3.10b) $[111]_r \rightarrow \mathbf{a}_m$	$\begin{pmatrix} 1 & 0 & 0 \\ 1 & \bar{1} & 1 \\ 1 & \bar{1} & \bar{1} \end{pmatrix}$	$\begin{pmatrix} 1 & 0 & 0 \\ 1 & \frac{\bar{1}}{2} & \frac{\bar{1}}{2} \\ 0 & \frac{1}{2} & \frac{\bar{1}}{2} \end{pmatrix}$	Rhombohedral space groups (*cf.* Section 4.3.5)
Primitive rhombohedral cell $\rightarrow$ A-centred monoclinic cell, unique axis **c**, cell choice 2 (Fig. 5.1.3.10b) $[111]_r \rightarrow \mathbf{a}_m$	$\begin{pmatrix} 1 & \bar{1} & \bar{1} \\ 1 & 0 & 0 \\ 1 & \bar{1} & 1 \end{pmatrix}$	$\begin{pmatrix} 0 & 1 & 0 \\ \frac{1}{2} & 1 & \frac{\bar{1}}{2} \\ \frac{\bar{1}}{2} & 0 & \frac{1}{2} \end{pmatrix}$	Rhombohedral space groups (*cf.* Section 4.3.5)

Table 5.1.3.1. *Selected 3 × 3 transformation matrices* **P** *and* **Q** = **P**⁻¹ (*cont.*)

Transformation	**P**	**Q** = **P**⁻¹	Crystal system
Primitive rhombohedral cell → A-centred monoclinic cell, unique axis **c**, cell choice 3 (Fig. 5.1.3.10b) [111]ᵣ → **a**ₘ	$\begin{pmatrix} 1 & \bar{1} & 1 \\ 1 & 1 & \bar{1} \\ 1 & 0 & 0 \end{pmatrix}$	$\begin{pmatrix} 0 & 0 & 1 \\ \frac{1}{2} & \frac{\bar{1}}{2} & 1 \\ \frac{1}{2} & \frac{\bar{1}}{2} & 0 \end{pmatrix}$	Rhombohedral space groups (*cf.* Section 4.3.5)

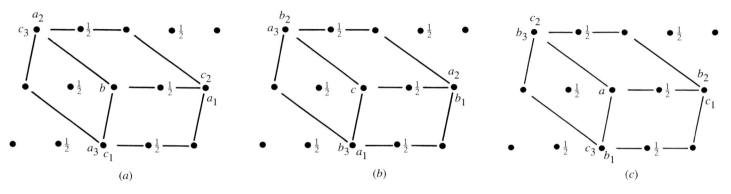

Fig. 5.1.3.2. Monoclinic centred lattice, projected along the unique axis. Origin for all cells is the same.

(a) Unique axis *b:*
    Cell choice 1: *C*-centred cell a_1, b, c_1.
    Cell choice 2: *A*-centred cell a_2, b, c_2.
    Cell choice 3: *I*-centred cell a_3, b, c_3.

(b) Unique axis *c:*
    Cell choice 1: *A*-centred cell a_1, b_1, c.
    Cell choice 2: *B*-centred cell a_2, b_2, c.
    Cell choice 3: *I*-centred cell a_3, b_3, c.

(c) Unique axis *a:*
    Cell choice 1: *B*-centred cell a, b_1, c_1.
    Cell choice 2: *C*-centred cell a, b_2, c_2.
    Cell choice 3: *I*-centred cell a, b_3, c_3.

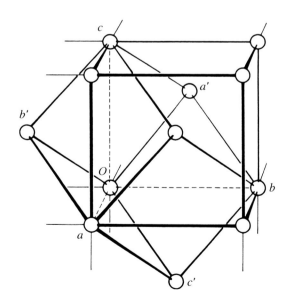

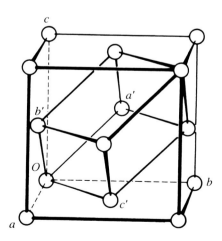

Fig. 5.1.3.3. Body-centred cell *I* with *a, b, c* and a corresponding primitive cell *P* with *a′, b′, c′*. Origin for both cells *O*. A cubic *I* cell with lattice constant a_c can be considered as a primitive rhombohedral cell with $a_r = a_c \frac{1}{2}\sqrt{3}$ and $\alpha = 109.47°$ (rhombohedral axes) or a triple hexagonal cell with $a_h = a_c\sqrt{2}$ and $c_h = a_c\frac{1}{2}\sqrt{3}$ (hexagonal axes).

Fig. 5.1.3.4. Face-centred cell *F* with *a, b, c* and a corresponding primitive cell *P* with *a′, b′, c′*. Origin for both cells *O*. A cubic *F* cell with lattice constant a_c can be considered as a primitive rhombohedral cell with $a_r = a_c\frac{1}{2}\sqrt{2}$ and $\alpha = 60°$ (rhombohedral axes) or a triple hexagonal cell with $a_h = a_c\frac{1}{2}\sqrt{2}$ and $c_h = a_c\sqrt{3}$ (hexagonal axes).

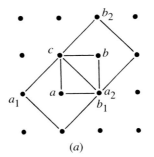 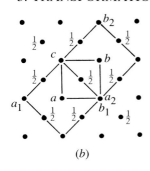

Fig. 5.1.3.5. Tetragonal lattices, projected along $[00\bar{1}]$. (a) Primitive cell P with a, b, c and the C-centred cells C_1 with a_1, b_1, c and C_2 with a_2, b_2, c. Origin for all three cells is the same. (b) Body-centred cell I with a, b, c and the F-centred cells F_1 with a_1, b_1, c and F_2 with a_2, b_2, c. Origin for all three cells is the same.

Fig. 5.1.3.7. Hexagonal lattice projected along $[00\bar{1}]$. Primitive hexagonal cell P with a, b, c and the three C-centred (orthohexagonal) cells a_1, b_1, c; a_2, b_2, c; a_3, b_3, c. Origin for all cells is the same.

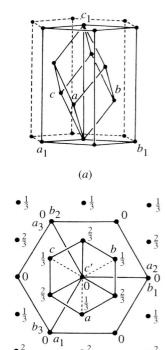 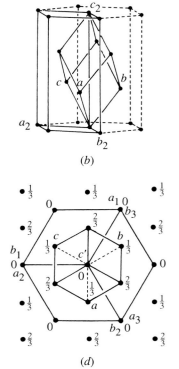

Fig. 5.1.3.6. Unit cells in the rhombohedral lattice: same origin for all cells. The basis of the rhombohedral cell is labelled a, b, c. Two settings of the triple hexagonal cell are possible with respect to a primitive rhombohedral cell: The *obverse setting* with the lattice points $0, 0, 0$; $\frac{2}{3}, \frac{1}{3}, \frac{1}{3}$; $\frac{1}{3}, \frac{2}{3}, \frac{2}{3}$ has been used in *International Tables* since 1952. Its general reflection condition is $-h + k + l = 3n$. The *reverse setting* with lattice points $0, 0, 0$; $\frac{1}{3}, \frac{2}{3}, \frac{1}{3}$; $\frac{2}{3}, \frac{1}{3}, \frac{2}{3}$ was used in the 1935 edition. Its general reflection condition is $h - k + l = 3n$. (a) Obverse setting of triple hexagonal cell a_1, b_1, c_1 in relation to the primitive rhombohedral cell a, b, c. (b) Reverse setting of triple hexagonal cell a_2, b_2, c_2 in relation to the primitive rhombohedral cell a, b, c. (c) Primitive rhombohedral cell (--- lower edges), a, b, c in relation to the three triple hexagonal cells in obverse setting a_1, b_1, c'; a_2, b_2, c'; a_3, b_3, c'. Projection along c'. (d) Primitive rhombohedral cell (- - - lower edges), a, b, c in relation to the three triple hexagonal cells in reverse setting a_1, b_1, c'; a_2, b_2, c'; a_3, b_3, c'. Projection along c'.

$$\begin{pmatrix} x' \\ y' \\ z' \\ 1 \end{pmatrix} = \mathbb{Q} \begin{pmatrix} x \\ y \\ z \\ 1 \end{pmatrix}$$

$$= \begin{pmatrix} Q_{11} & Q_{12} & Q_{13} & q_1 \\ Q_{21} & Q_{22} & Q_{23} & q_2 \\ Q_{31} & Q_{32} & Q_{33} & q_3 \\ 0 & 0 & 0 & 1 \end{pmatrix} \begin{pmatrix} x \\ y \\ z \\ 1 \end{pmatrix}$$

$$= \begin{pmatrix} Q_{11}x + Q_{12}y + Q_{13}z + q_1 \\ Q_{21}x + Q_{22}y + Q_{23}z + q_2 \\ Q_{31}x + Q_{32}y + Q_{33}z + q_3 \\ 1 \end{pmatrix}.$$

The inverse of the augmented matrix $\mathbb{Q}$ is the augmented matrix $\mathbb{P}$ which contains the matrices P and p, specifically,

$$\mathbb{P} = \mathbb{Q}^{-1} = \begin{pmatrix} P & p \\ o & 1 \end{pmatrix} = \begin{pmatrix} Q^{-1} & -Q^{-1}q \\ o & 1 \end{pmatrix}.$$

The advantage of the use of (4×4) matrices is that a sequence of affine transformations corresponds to the product of the correspond-

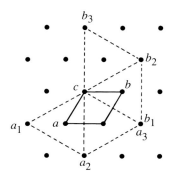

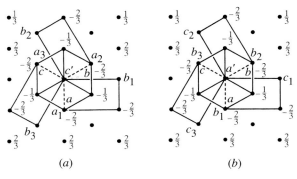

Fig. 5.1.3.8. Hexagonal lattice projected along $[00\bar{1}]$. Primitive hexagonal cell P with a, b, c and the three triple hexagonal cells H with a_1, b_1, c; a_2, b_2, c; a_3, b_3, c. Origin for all cells is the same.

Fig. 5.1.3.10. Rhombohedral lattice with primitive rhombohedral cell a, b, c and the three centred monoclinic cells. (a) C-centred cells C_1 with a_1, b_1, c'; C_2 with a_2, b_2, c'; and C_3 with a_3, b_3, c'. The unique monoclinic axes are b_1, b_2 and b_3, respectively. Origin for all four cells is the same. (b) A-centred cells A_1 with a', b_1, c_1; A_2 with a', b_2, c_2; and A_3 with a', b_3, c_3. The unique monoclinic axes are c_1, c_2 and c_3, respectively. Origin for all four cells is the same.

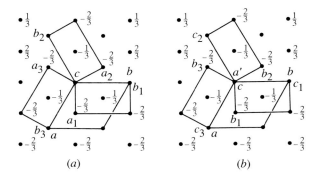

Fig. 5.1.3.9. Rhombohedral lattice with a triple hexagonal unit cell a, b, c in obverse setting (i.e. unit cell a_1, b_1, c in Fig. 5.1.3.6c) and the three centred monoclinic cells. (a) C-centred cells C_1 with a_1, b_1, c; C_2 with a_2, b_2, c; and C_3 with a_3, b_3, c. The unique monoclinic axes are b_1, b_2 and b_3, respectively. Origin for all four cells is the same. (b) A-centred cells A_1 with a', b_1, c_1; A_2 with a', b_2, c_2; and A_3 with a', b_3, c_3. The unique monoclinic axes are c_1, c_2 and c_3, respectively. Origin for all four cells is the same.

ing matrices. However, the order of the factors in the product must be observed. If $\mathbb{Q}$ is the product of n transformation matrices $\mathbb{Q}_i$,

$$\mathbb{Q} = \mathbb{Q}_n \ldots \mathbb{Q}_2 \mathbb{Q}_1,$$

the sequence of the corresponding inverse matrices $\mathbb{P}_i$ is reversed in the product

$$\mathbb{P} = \mathbb{P}_1 \mathbb{P}_2 \ldots \mathbb{P}_n.$$

The following items are also affected by a transformation:

(i) The *metric matrix of direct lattice* $\mathbf{G}$ [more exactly: the matrix of geometrical coefficients (metric tensor)] is transformed by the matrix $\mathbf{P}$ as follows:

$$\mathbf{G}' = \mathbf{P}^t \mathbf{G} \mathbf{P}$$

with $\mathbf{P}^t$ the transposed matrix of $\mathbf{P}$, i.e. rows and columns of $\mathbf{P}$ are interchanged. Specifically,

$$
\begin{aligned}
\mathbf{G}' &= \begin{pmatrix} \mathbf{a}' \cdot \mathbf{a}' & \mathbf{a}' \cdot \mathbf{b}' & \mathbf{a}' \cdot \mathbf{c}' \\ \mathbf{b}' \cdot \mathbf{a}' & \mathbf{b}' \cdot \mathbf{b}' & \mathbf{b}' \cdot \mathbf{c}' \\ \mathbf{c}' \cdot \mathbf{a}' & \mathbf{c}' \cdot \mathbf{b}' & \mathbf{c}' \cdot \mathbf{c}' \end{pmatrix} \\
&= \begin{pmatrix} P_{11} & P_{21} & P_{31} \\ P_{12} & P_{22} & P_{32} \\ P_{13} & P_{23} & P_{33} \end{pmatrix} \begin{pmatrix} \mathbf{a} \cdot \mathbf{a} & \mathbf{a} \cdot \mathbf{b} & \mathbf{a} \cdot \mathbf{c} \\ \mathbf{b} \cdot \mathbf{a} & \mathbf{b} \cdot \mathbf{b} & \mathbf{b} \cdot \mathbf{c} \\ \mathbf{c} \cdot \mathbf{a} & \mathbf{c} \cdot \mathbf{b} & \mathbf{c} \cdot \mathbf{c} \end{pmatrix} \\
&\quad \times \begin{pmatrix} P_{11} & P_{12} & P_{13} \\ P_{21} & P_{22} & P_{23} \\ P_{31} & P_{32} & P_{33} \end{pmatrix}.
\end{aligned}
$$

(ii) The *metric matrix of reciprocal lattice* $\mathbf{G}^*$ [more exactly: the matrix of geometrical coefficients (metric tensor)] is transformed by

$$\mathbf{G}^{*\prime} = \mathbf{Q} \mathbf{G}^* \mathbf{Q}^t.$$

Here, the transposed matrix $\mathbf{Q}^t$ is on the right-hand side of $\mathbf{G}^*$.

(iii) The *volume of the unit cell* V changes with the transformation. The volume of the new unit cell V' is obtained by

$$V' = \det(\mathbf{P})V = \begin{vmatrix} P_{11} & P_{12} & P_{13} \\ P_{21} & P_{22} & P_{23} \\ P_{31} & P_{32} & P_{33} \end{vmatrix} V$$

with $\det(\mathbf{P})$ the determinant of the matrix $\mathbf{P}$. The corresponding equation for the volume of the unit cell in reciprocal space V^* is

$$V^{*\prime} = \det(\mathbf{Q})V^*.$$

Matrices $\mathbf{P}$ and $\mathbf{Q}$ that frequently occur in crystallography are listed in Table 5.1.3.1.

5.2. Transformations of symmetry operations (motions)

By H. Arnold

5.2.1. Transformations

Symmetry operations are transformations in which the coordinate system, *i.e.* the basis vectors **a**, **b**, **c** and the origin O, are considered to be at rest, whereas the object is mapped onto itself. This can be visualized as a 'motion' of an object in such a way that the object before and after the 'motion' cannot be distinguished.

A symmetry operation W transforms every point X with the coordinates x, y, z to a symmetrically equivalent point $\tilde{X}$ with the coordinates $\tilde{x}, \tilde{y}, \tilde{z}$. In matrix notation, this transformation is performed by

$$
\begin{pmatrix} \tilde{x} \\ \tilde{y} \\ \tilde{z} \end{pmatrix} = \begin{pmatrix} W_{11} & W_{12} & W_{13} \\ W_{21} & W_{22} & W_{23} \\ W_{31} & W_{32} & W_{33} \end{pmatrix} \begin{pmatrix} x \\ y \\ z \end{pmatrix} + \begin{pmatrix} w_1 \\ w_2 \\ w_3 \end{pmatrix}
$$
$$
= \begin{pmatrix} W_{11}x + W_{12}y + W_{13}z + w_1 \\ W_{21}x + W_{22}y + W_{23}z + w_2 \\ W_{31}x + W_{32}y + W_{33}z + w_3 \end{pmatrix}.
$$

The (3×3) matrix $\boldsymbol{W}$ is the rotation part and the (3×1) column matrix $\boldsymbol{w}$ the translation part of the symmetry operation W. The pair $(\boldsymbol{W}, \boldsymbol{w})$ characterizes the operation uniquely. Matrices $\boldsymbol{W}$ for point-group operations are given in Tables 11.2.2.1 and 11.2.2.2.

Again, we can introduce the augmented (4×4) matrix (*cf.* Chapter 8.1)

$$
\mathsf{W} = \begin{pmatrix} \boldsymbol{W} & \boldsymbol{w} \\ \boldsymbol{o} & 1 \end{pmatrix} = \begin{pmatrix} W_{11} & W_{12} & W_{13} & w_1 \\ W_{21} & W_{22} & W_{23} & w_2 \\ W_{31} & W_{32} & W_{33} & w_3 \\ 0 & 0 & 0 & 1 \end{pmatrix}.
$$

The coordinates $\tilde{x}, \tilde{y}, \tilde{z}$ of the point $\tilde{X}$, symmetrically equivalent to X with the coordinates x, y, z, are obtained by

$$
\begin{pmatrix} \tilde{x} \\ \tilde{y} \\ \tilde{z} \\ 1 \end{pmatrix} = \begin{pmatrix} W_{11} & W_{12} & W_{13} & w_1 \\ W_{21} & W_{22} & W_{23} & w_2 \\ W_{31} & W_{32} & W_{33} & w_3 \\ 0 & 0 & 0 & 1 \end{pmatrix} \begin{pmatrix} x \\ y \\ z \\ 1 \end{pmatrix}
$$
$$
= \begin{pmatrix} W_{11}x + W_{12}y + W_{13}z + w_1 \\ W_{21}x + W_{22}y + W_{23}z + w_2 \\ W_{31}x + W_{32}y + W_{33}z + w_3 \\ 1 \end{pmatrix},
$$

or, in short notation,

$$
\tilde{\mathsf{x}} = \mathsf{W}\mathsf{x}.
$$

A sequence of symmetry operations can be obtained as a product of (4×4) matrices W.

An affine transformation of the coordinate system transforms the coordinates x of the starting point

$$
\mathsf{x}' = \mathbb{Q}\mathsf{x}
$$

as well as the coordinates $\tilde{\mathsf{x}}$ of a symmetrically equivalent point

$$
\tilde{\mathsf{x}}' = \mathbb{Q}\tilde{\mathsf{x}}
$$
$$
= \mathbb{Q}\mathsf{W}\mathsf{x}
$$
$$
= \mathbb{Q}\mathsf{W}\mathbb{P}\mathbb{Q}\mathsf{x} \quad \text{(with } \mathbb{P} = \mathbb{Q}^{-1}\text{)}
$$
$$
= \mathbb{Q}\mathsf{W}\mathbb{P}\mathsf{x}'.
$$

Thus, the affine transformation transforms also the symmetry-operation matrix W and the new matrix W' is obtained by

$$
\mathsf{W}' = \mathbb{Q}\mathsf{W}\mathbb{P}.
$$

Example

Space group $P4/n$ (85) is listed in the space-group tables with two origins; origin choice 1 with $\bar{4}$, origin choice 2 with $\bar{1}$ as point symmetry of the origin. How does the matrix W of the symmetry operation $\bar{4}^+\ 0, 0, z;\ 0, 0, 0$ of origin choice 1 transform to the matrix W' of symmetry operation $\bar{4}^+\ \frac{1}{4}, -\frac{1}{4}, z;\ \frac{1}{4}, -\frac{1}{4}, 0$ of origin choice 2?

In the space-group tables, origin choice 1, the transformed coordinates $\tilde{x}, \tilde{y},\ \tilde{z} = y, \bar{x}, \bar{z}$ are listed. The translation part is zero, *i.e.* $\boldsymbol{w} = (0/0/0)$. In Table 11.2.2.1, the matrix $\boldsymbol{W}$ can be found. Thus, the (4×4) matrix W is obtained:

$$
\mathsf{W} = \begin{pmatrix} \boldsymbol{W} & \boldsymbol{w} \\ \boldsymbol{o} & 1 \end{pmatrix} = \begin{pmatrix} 0 & 1 & 0 & 0 \\ \bar{1} & 0 & 0 & 0 \\ 0 & 0 & \bar{1} & 0 \\ 0 & 0 & 0 & 1 \end{pmatrix}.
$$

The transformation to origin choice 2 is accomplished by a shift vector **p** with components $\frac{1}{4}, -\frac{1}{4}, 0$. Since this is a pure shift, the matrices $\boldsymbol{P}$ and $\boldsymbol{Q}$ are the unit matrix $\boldsymbol{I}$. Now the shift vector $\boldsymbol{q}$ is derived: $\boldsymbol{q} = -\boldsymbol{P}^{-1}\boldsymbol{p} = -\boldsymbol{I}\boldsymbol{p} = -\boldsymbol{p}$. Thus, the matrices $\mathbb{P}$ and $\mathbb{Q}$ are

$$
\mathbb{P} = \begin{pmatrix} 1 & 0 & 0 & \frac{1}{4} \\ 0 & 1 & 0 & \frac{\bar{1}}{4} \\ 0 & 0 & 1 & 0 \\ 0 & 0 & 0 & 1 \end{pmatrix}, \quad \mathbb{Q} = \begin{pmatrix} 1 & 0 & 0 & \frac{\bar{1}}{4} \\ 0 & 1 & 0 & \frac{1}{4} \\ 0 & 0 & 1 & 0 \\ 0 & 0 & 0 & 1 \end{pmatrix}.
$$

By matrix multiplication, the new matrix W' is obtained:

$$
\mathsf{W}' = \mathbb{Q}\mathsf{W}\mathbb{P} = \begin{pmatrix} 0 & 1 & 0 & \frac{\bar{1}}{2} \\ \bar{1} & 0 & 0 & 0 \\ 0 & 0 & \bar{1} & 0 \\ 0 & 0 & 0 & 1 \end{pmatrix}.
$$

If the matrix W' is applied to x', y', z', the coordinates of the starting point in the new coordinate system, we obtain the transformed coordinates $\tilde{x}', \tilde{y}', \tilde{z}'$,

$$
\begin{pmatrix} \tilde{x}' \\ \tilde{y}' \\ \tilde{z}' \\ 1 \end{pmatrix} = \begin{pmatrix} 0 & 1 & 0 & \frac{\bar{1}}{2} \\ \bar{1} & 0 & 0 & 0 \\ 0 & 0 & \bar{1} & 0 \\ 0 & 0 & 0 & 1 \end{pmatrix} \begin{pmatrix} x' \\ y' \\ z' \\ 1 \end{pmatrix} = \begin{pmatrix} y' - \frac{1}{2} \\ \bar{x}' \\ \bar{z}' \\ 1 \end{pmatrix}.
$$

By adding a lattice translation **a**, the transformed coordinates $y + \frac{1}{2}, \bar{x}, \bar{z}$ are obtained as listed in the space-group tables for origin choice 2.

5.2.2. Invariants

A crystal structure and its physical properties are independent of the choice of the unit cell. This implies that invariants occur, *i.e.* quantities which have the same values before and after the transformation. Only some important invariants are considered in this section. Invariants of higher order (tensors) are treated by Altmann & Herzig (1994), second cumulant tensors, *i.e.* anisotropic temperature factors, are given in *International Tables for Crystallography* (2004), Vol. C.

The *orthogonality of the basis vectors* **a**, **b**, **c** of direct space and the basis vectors **a***, **b***, **c*** of reciprocal space,

$$\begin{pmatrix} \mathbf{a}^* \\ \mathbf{b}^* \\ \mathbf{c}^* \end{pmatrix} (\mathbf{a}, \mathbf{b}, \mathbf{c}) = \begin{pmatrix} \mathbf{a}^* \cdot \mathbf{a} & \mathbf{a}^* \cdot \mathbf{b} & \mathbf{a}^* \cdot \mathbf{c} \\ \mathbf{b}^* \cdot \mathbf{a} & \mathbf{b}^* \cdot \mathbf{b} & \mathbf{b}^* \cdot \mathbf{c} \\ \mathbf{c}^* \cdot \mathbf{a} & \mathbf{c}^* \cdot \mathbf{b} & \mathbf{c}^* \cdot \mathbf{c} \end{pmatrix} = \mathbf{I},$$

is invariant under a general (affine) transformation. Since both sets of basis vectors are transformed, **a*** is always perpendicular to the plane defined by **b** and **c** and **a***$'$ perpendicular to **b**$'$ and **c**$'$ *etc.*

5.2.2.1. *Position vector*

The position vector **r** in direct space,

$$\mathbf{r} = (\mathbf{a}, \mathbf{b}, \mathbf{c}) \begin{pmatrix} x \\ y \\ z \end{pmatrix} = x\mathbf{a} + y\mathbf{b} + z\mathbf{c},$$

is invariant if the origin of the coordinate system is not changed in the transformation (see example in Section 5.1.3).

5.2.2.2. *Modulus of position vector*

The modulus r of the position vector **r** gives the distance of the point x, y, z from the origin. Its square is obtained by the scalar product

$$\mathbf{r}^t \cdot \mathbf{r} = r^2 = (x, y, z) \begin{pmatrix} \mathbf{a} \\ \mathbf{b} \\ \mathbf{c} \end{pmatrix} (\mathbf{a}, \mathbf{b}, \mathbf{c}) \begin{pmatrix} x \\ y \\ z \end{pmatrix}$$

$$= (x, y, z) \mathbf{G} \begin{pmatrix} x \\ y \\ z \end{pmatrix}$$

$$= x^2 a^2 + y^2 b^2 + z^2 c^2 + 2yzbc \cos\alpha$$
$$+ 2xzac \cos\beta + 2xyab \cos\gamma,$$

with $\mathbf{r}^t$ the transposed representation of **r**; a, b, c the moduli of the basis vectors **a**, **b**, **c** (lattice parameters); **G** the metric matrix of direct space; and α, β, γ the angles of the unit cell.

The same considerations apply to the vector **r*** in reciprocal space and its modulus r^*. Here, **G*** is applied. Note that **r*** and r^* are independent of the choice of the origin in direct space.

5.2.2.3. *Metric matrix*

The metric matrix **G** of the unit cell in the direct lattice

$$\mathbf{G} = \begin{pmatrix} \mathbf{a} \cdot \mathbf{a} & \mathbf{a} \cdot \mathbf{b} & \mathbf{a} \cdot \mathbf{c} \\ \mathbf{b} \cdot \mathbf{a} & \mathbf{b} \cdot \mathbf{b} & \mathbf{b} \cdot \mathbf{c} \\ \mathbf{c} \cdot \mathbf{a} & \mathbf{c} \cdot \mathbf{b} & \mathbf{c} \cdot \mathbf{c} \end{pmatrix} = \begin{pmatrix} aa & ab\cos\gamma & ac\cos\beta \\ ba\cos\gamma & bb & bc\cos\alpha \\ ca\cos\beta & cb\cos\alpha & cc \end{pmatrix}$$

changes under a linear transformation, but **G** is invariant under a symmetry operation of the lattice. The volume of the unit cell V is obtained by

$$V^2 = \det(\mathbf{G}).$$

The same considerations apply to the metric matrix **G*** of the unit cell in the reciprocal lattice and the volume V^* of the reciprocal-lattice unit cell. Thus, there are two invariants under an affine transformation, the product

$$VV^* = 1$$

and the product

$$GG^* = I.$$

5.2.2.4. *Scalar product*

The scalar product

$$\mathbf{r}^* \cdot \mathbf{r} = hx + ky + lz$$

of the vector **r*** in reciprocal space with the vector **r** in direct space is invariant under a linear transformation but not under a shift of origin in direct space.

A vector **r** in direct space can also be represented as a product of augmented matrices:

$$\mathbf{r} = (\mathbf{a}, \mathbf{b}, \mathbf{c}, 0) \begin{pmatrix} x \\ y \\ z \\ 1 \end{pmatrix} = x\mathbf{a} + y\mathbf{b} + z\mathbf{c}.$$

As stated above, the basis vectors are transformed only by the linear part, even in the case of a general affine transformation. Thus, the transformed position vector **r**$'$ is obtained by

$$\mathbf{r}' = (\mathbf{a}, \mathbf{b}, \mathbf{c}, 0) \begin{pmatrix} \mathbf{P} & \mathbf{o}^t \\ \mathbf{o} & 1 \end{pmatrix} \begin{pmatrix} \mathbf{Q} & \mathbf{q} \\ \mathbf{o} & 1 \end{pmatrix} \begin{pmatrix} x \\ y \\ z \\ 1 \end{pmatrix}.$$

The shift **p** is set to zero. The shift of origin is contained in the matrix $\mathbb{Q}$ only.

Similarly, a vector in reciprocal space can be represented by

$$\mathbf{r}^* = (h, k, l, 1) \begin{pmatrix} \mathbf{a}^* \\ \mathbf{b}^* \\ \mathbf{c}^* \\ 0 \end{pmatrix} = h\mathbf{a}^* + k\mathbf{b}^* + l\mathbf{c}^*.$$

The coordinates h, k, l in reciprocal space transform also only linearly. Thus,

$$\mathbf{r}^{*\prime} = (h, k, l, 1) \begin{pmatrix} \mathbf{P} & \mathbf{o}^t \\ \mathbf{o} & 1 \end{pmatrix} \begin{pmatrix} \mathbf{Q} & \mathbf{q} \\ \mathbf{o} & 1 \end{pmatrix} \begin{pmatrix} \mathbf{a}^* \\ \mathbf{b}^* \\ \mathbf{c}^* \\ 0 \end{pmatrix}.$$

The reader can see immediately that the scalar product $\mathbf{r}^* \cdot \mathbf{r}$ transforms correctly.

5.2.3. Example: low cristobalite and high cristobalite

The positions of the silicon atoms in the low-cristobalite structure (Nieuwenkamp, 1935) are compared with those of the high-cristobalite structure (Wyckoff, 1925; *cf.* Megaw, 1973). At low temperatures, the space group is $P4_1 2_1 2$ (92). The four silicon atoms are located in Wyckoff position $4(a)$ ..2 with the coordinates $x, x, 0; \bar{x}, \bar{x}, \frac{1}{2}; \frac{1}{2} - x, \frac{1}{2} + x, \frac{1}{4}; \frac{1}{2} + x, \frac{1}{2} - x, \frac{3}{4}; x = 0.300$. During the phase transition, the tetragonal structure is transformed into a cubic one with space group $Fd\bar{3}m$ (227). It is listed in the space-group tables with two different origins. We use 'Origin choice 1' with point symmetry $\bar{4}3m$ at the origin. The silicon atoms occupy the position $8(a)$ $\bar{4}3m$ with the coordinates $0, 0, 0; \frac{1}{4}, \frac{1}{4}, \frac{1}{4}$ and those related by the face-centring translations. In the diamond structure, the carbon atoms occupy the same position.

In order to compare the two structures, the conventional P cell of space group $P4_1 2_1 2$ (92) is transformed to an unconventional C cell (*cf.* Section 4.3.4), which corresponds to the F cell of $Fd\bar{3}m$ (227). The P and the C cells are shown in Fig. 5.2.3.1. The coordinate system **a**$'$, **b**$'$, **c**$'$ with origin O' of the C cell is obtained from that of the P cell, origin O, by the linear transformation

$$\mathbf{a}' = \mathbf{a} + \mathbf{b}, \qquad \mathbf{b}' = -\mathbf{a} + \mathbf{b}, \qquad \mathbf{c}' = \mathbf{c}$$

and the shift

$$p = \tfrac{1}{4}\mathbf{a} + \tfrac{1}{4}\mathbf{b}.$$

The matrices $\boldsymbol{P}, \boldsymbol{p}$ and $\mathbb{P}$ are thus given by

$$\boldsymbol{P} = \begin{pmatrix} 1 & \bar{1} & 0 \\ 1 & 1 & 0 \\ 0 & 0 & 1 \end{pmatrix}, \quad \boldsymbol{p} = \begin{pmatrix} \tfrac{1}{4} \\ \tfrac{1}{4} \\ 0 \end{pmatrix}, \quad \mathbb{P} = \begin{pmatrix} 1 & \bar{1} & 0 & \tfrac{1}{4} \\ 1 & 1 & 0 & \tfrac{1}{4} \\ 0 & 0 & 1 & 0 \\ 0 & 0 & 0 & 1 \end{pmatrix}.$$

From Fig. 5.2.3.1, we derive also the inverse transformation

$$\mathbf{a} = \tfrac{1}{2}\mathbf{a}' - \tfrac{1}{2}\mathbf{b}', \quad \mathbf{b} = \tfrac{1}{2}\mathbf{a}' + \tfrac{1}{2}\mathbf{b}', \quad \mathbf{c} = \mathbf{c}', \quad \mathbf{q} = -\tfrac{1}{4}\mathbf{a}'.$$

Thus, the matrices $\boldsymbol{Q}, \boldsymbol{q}$ and $\mathbb{Q}$ are

$$\boldsymbol{Q} = \boldsymbol{P}^{-1} = \begin{pmatrix} \tfrac{1}{2} & \tfrac{1}{2} & 0 \\ \bar{\tfrac{1}{2}} & \tfrac{1}{2} & 0 \\ 0 & 0 & 1 \end{pmatrix}, \quad \boldsymbol{q} = -\boldsymbol{P}^{-1}\boldsymbol{p} = \begin{pmatrix} \bar{\tfrac{1}{4}} \\ 0 \\ 0 \end{pmatrix},$$

$$\mathbb{Q} = \mathbb{P}^{-1} = \begin{pmatrix} \tfrac{1}{2} & \tfrac{1}{2} & 0 & \bar{\tfrac{1}{4}} \\ \bar{\tfrac{1}{2}} & \tfrac{1}{2} & 0 & 0 \\ 0 & 0 & 1 & 0 \\ 0 & 0 & 0 & 1 \end{pmatrix}.$$

The coordinates x, y, z of points in the P cell are transformed by $\mathbb{Q}$:

$$\begin{pmatrix} x' \\ y' \\ z' \\ 1 \end{pmatrix} = \begin{pmatrix} \tfrac{1}{2} & \tfrac{1}{2} & 0 & \bar{\tfrac{1}{4}} \\ \bar{\tfrac{1}{2}} & \tfrac{1}{2} & 0 & 0 \\ 0 & 0 & 1 & 0 \\ 0 & 0 & 0 & 1 \end{pmatrix} \begin{pmatrix} x \\ y \\ z \\ 1 \end{pmatrix} = \begin{pmatrix} \tfrac{1}{2}(x+y) - \tfrac{1}{4} \\ \tfrac{1}{2}(-x+y) \\ z \\ 1 \end{pmatrix}.$$

The coordinate triplets of the four silicon positions in the P cell are $0.300, 0.300, 0$; $0.700, 0.700, \tfrac{1}{2}$; $0.200, 0.800, \tfrac{1}{4}$; $0.800, 0.200, \tfrac{3}{4}$. Four triplets in the C cell are obtained by inserting these values into the equation just derived. The new coordinates are $0.050, 0, 0$; $0.450, 0, \tfrac{1}{2}$; $0.250, 0.300, \tfrac{1}{4}$; $0.250, -0.300, \tfrac{3}{4}$. A set of four further points is obtained by adding the centring translation $\tfrac{1}{2}, \tfrac{1}{2}, 0$ to these coordinates.

The indices h, k, l are transformed by the matrix $\boldsymbol{P}$:

$$(h', k', l') = (h, k, l) \begin{pmatrix} 1 & \bar{1} & 0 \\ 1 & 1 & 0 \\ 0 & 0 & 1 \end{pmatrix} = (h+k, -h+k, l),$$

i.e. the reflections with the indices h, k, l of the P cell become reflections $h+k, -h+k, l$ of the C cell.

The symmetry operations of space group $P4_12_12$ are listed in the space-group tables for the P cell as follows:

(1) x, y, z;   (2) $\bar{x}, \bar{y}, \tfrac{1}{2} + z$;

(3) $\tfrac{1}{2} - y, \tfrac{1}{2} + x, \tfrac{1}{4} + z$;   (4) $\tfrac{1}{2} + y, \tfrac{1}{2} - x, \tfrac{3}{4} + z$;

(5) $\tfrac{1}{2} - x, \tfrac{1}{2} + y, \tfrac{1}{4} - z$;   (6) $\tfrac{1}{2} + x, \tfrac{1}{2} - y, \tfrac{3}{4} - z$;

(7) $y, x, \bar{z}$;   (8) $\bar{y}, \bar{x}, \tfrac{1}{2} - z$.

The corresponding matrices $\mathbb{W}$ are

$$(1)\begin{pmatrix} 1 & 0 & 0 & 0 \\ 0 & 1 & 0 & 0 \\ 0 & 0 & 1 & 0 \\ 0 & 0 & 0 & 1 \end{pmatrix}; \quad (2)\begin{pmatrix} \bar{1} & 0 & 0 & 0 \\ 0 & \bar{1} & 0 & 0 \\ 0 & 0 & 1 & \tfrac{1}{2} \\ 0 & 0 & 0 & 1 \end{pmatrix}; \quad (3)\begin{pmatrix} 0 & \bar{1} & 0 & \tfrac{1}{2} \\ 1 & 0 & 0 & \tfrac{1}{2} \\ 0 & 0 & 1 & \tfrac{1}{4} \\ 0 & 0 & 0 & 1 \end{pmatrix};$$

$$(4)\begin{pmatrix} 0 & 1 & 0 & \tfrac{1}{2} \\ \bar{1} & 0 & 0 & \tfrac{1}{2} \\ 0 & 0 & 1 & \tfrac{3}{4} \\ 0 & 0 & 0 & 1 \end{pmatrix}; \quad (5)\begin{pmatrix} \bar{1} & 0 & 0 & \tfrac{1}{2} \\ 0 & 1 & 0 & \tfrac{1}{2} \\ 0 & 0 & \bar{1} & \tfrac{1}{4} \\ 0 & 0 & 0 & 1 \end{pmatrix}; \quad (6)\begin{pmatrix} 1 & 0 & 0 & \tfrac{1}{2} \\ 0 & \bar{1} & 0 & \tfrac{1}{2} \\ 0 & 0 & \bar{1} & \tfrac{3}{4} \\ 0 & 0 & 0 & 1 \end{pmatrix};$$

$$(7)\begin{pmatrix} 0 & 1 & 0 & 0 \\ 1 & 0 & 0 & 0 \\ 0 & 0 & \bar{1} & 0 \\ 0 & 0 & 0 & 1 \end{pmatrix}; \quad (8)\begin{pmatrix} 0 & \bar{1} & 0 & 0 \\ \bar{1} & 0 & 0 & 0 \\ 0 & 0 & \bar{1} & \tfrac{1}{2} \\ 0 & 0 & 0 & 1 \end{pmatrix}.$$

These matrices of the P cell are transformed to the matrices $\mathbb{W}'$ of the C cell by

$$\mathbb{W}' = \mathbb{Q}\mathbb{W}\mathbb{P}.$$

For matrix (2), for example, this results in

$$\mathbb{W}' = \begin{pmatrix} \tfrac{1}{2} & \tfrac{1}{2} & 0 & \bar{\tfrac{1}{4}} \\ \bar{\tfrac{1}{2}} & \tfrac{1}{2} & 0 & 0 \\ 0 & 0 & 1 & 0 \\ 0 & 0 & 0 & 1 \end{pmatrix} \begin{pmatrix} \bar{1} & 0 & 0 & 0 \\ 0 & \bar{1} & 0 & 0 \\ 0 & 0 & 1 & \tfrac{1}{2} \\ 0 & 0 & 0 & 1 \end{pmatrix} \begin{pmatrix} 1 & \bar{1} & 0 & \tfrac{1}{4} \\ 1 & 1 & 0 & \tfrac{1}{4} \\ 0 & 0 & 1 & 0 \\ 0 & 0 & 0 & 1 \end{pmatrix}$$

$$= \begin{pmatrix} \bar{1} & 0 & 0 & \tfrac{1}{2} \\ 0 & \bar{1} & 0 & 0 \\ 0 & 0 & 1 & \tfrac{1}{2} \\ 0 & 0 & 0 & 1 \end{pmatrix}.$$

The eight transformed matrices $\mathbb{W}'$, derived in this way, are

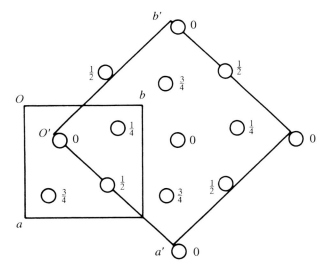

Fig. 5.2.3.1. Positions of silicon atoms in the low-cristobalite structure, projected along $[00\bar{1}]$. Primitive tetragonal cell a, b, c; C-centred tetragonal cell a', b', c'. Shift of origin from O to O' by the vector $p = \tfrac{1}{4}\mathbf{a} + \tfrac{1}{4}\mathbf{b}$.

$$(1) \begin{pmatrix} 1 & 0 & 0 & 0 \\ 0 & 1 & 0 & 0 \\ 0 & 0 & 1 & 0 \\ 0 & 0 & 0 & 1 \end{pmatrix}; \quad (2) \begin{pmatrix} \bar{1} & 0 & 0 & \frac{1}{2} \\ 0 & \bar{1} & 0 & 0 \\ 0 & 0 & 1 & \frac{1}{2} \\ 0 & 0 & 0 & 1 \end{pmatrix}; \quad (3) \begin{pmatrix} 0 & \bar{1} & 0 & \frac{1}{4} \\ 1 & 0 & 0 & \frac{1}{4} \\ 0 & 0 & 1 & \frac{1}{4} \\ 0 & 0 & 0 & 1 \end{pmatrix};$$

$$(4) \begin{pmatrix} 0 & 1 & 0 & \frac{1}{4} \\ \bar{1} & 0 & 0 & \frac{\bar{1}}{4} \\ 0 & 0 & 1 & \frac{3}{4} \\ 0 & 0 & 0 & 1 \end{pmatrix}; \quad (5) \begin{pmatrix} 0 & 1 & 0 & \frac{1}{4} \\ 1 & 0 & 0 & \frac{1}{4} \\ 0 & 0 & \bar{1} & \frac{1}{4} \\ 0 & 0 & 0 & 1 \end{pmatrix}; \quad (6) \begin{pmatrix} 0 & \bar{1} & 0 & \frac{1}{4} \\ \bar{1} & 0 & 0 & \frac{\bar{1}}{4} \\ 0 & 0 & \bar{1} & \frac{3}{4} \\ 0 & 0 & 0 & 1 \end{pmatrix};$$

$$(7) \begin{pmatrix} 1 & 0 & 0 & 0 \\ 0 & \bar{1} & 0 & 0 \\ 0 & 0 & \bar{1} & 0 \\ 0 & 0 & 0 & 1 \end{pmatrix}; \quad (8) \begin{pmatrix} \bar{1} & 0 & 0 & \frac{1}{2} \\ 0 & 1 & 0 & 0 \\ 0 & 0 & \bar{1} & \frac{1}{2} \\ 0 & 0 & 0 & 1 \end{pmatrix}.$$

Another set of eight matrices is obtained by adding the C-centring translation $\frac{1}{2}, \frac{1}{2}, 0$ to the $\boldsymbol{w}$'s.

From these matrices, one obtains the coordinates of the general position in the C cell, for instance from matrix (2)

$$\begin{pmatrix} \tilde{x} \\ \tilde{y} \\ \tilde{z} \\ 1 \end{pmatrix} = \begin{pmatrix} \bar{1} & 0 & 0 & \frac{\bar{1}}{2} \\ 0 & \bar{1} & 0 & 0 \\ 0 & 0 & 1 & \frac{1}{2} \\ 0 & 0 & 0 & 1 \end{pmatrix} \begin{pmatrix} x \\ y \\ z \\ 1 \end{pmatrix} = \begin{pmatrix} -x - \frac{1}{2} \\ -y \\ z + \frac{1}{2} \\ 1 \end{pmatrix}.$$

The eight points obtained by the eight matrices $\mathbb{W}'$ are

(1) $x, y, z;$ (2) $-\frac{1}{2} - x, \bar{y}, \frac{1}{2} + z;$

(3) $\frac{1}{4} - y, \frac{1}{4} + x, \frac{1}{4} + z;$ (4) $\frac{1}{4} + y, -\frac{1}{4} - x, \frac{3}{4} + z;$

(5) $\frac{1}{4} + y, \frac{1}{4} + x, \frac{1}{4} - z;$ (6) $\frac{1}{4} - y, -\frac{1}{4} - x, \frac{3}{4} - z;$

(7) $x, \bar{y}, \bar{z};$ (8) $-\frac{1}{2} - x, y, \frac{1}{2} - z.$

The other set of eight points is obtained by adding $\frac{1}{2}, \frac{1}{2}, 0$.

In space group $P4_1 2_1 2$, the silicon atoms are in special position $4(a)$ ..2 with the coordinates $x, x, 0$. Transformed into the C cell, the position becomes

$$(0, 0, 0) + \quad (\tfrac{1}{2}, \tfrac{1}{2}, 0) +$$
$$x, 0, 0; \quad \tfrac{1}{2} - x, 0, \tfrac{1}{2}; \quad \tfrac{1}{4}, \tfrac{1}{4} + x, \tfrac{1}{4}; \quad \tfrac{1}{4}, \tfrac{3}{4} - x, \tfrac{3}{4}.$$

The parameter $x = 0.300$ of the P cell has changed to $x = 0.050$ in the C cell. For $x = 0$, the special position of the C cell assumes the same coordinate triplets as Wyckoff position $8(a)$ $\bar{4}3m$ in space group $Fd\bar{3}m$ (227), *i.e.* this change of the x parameter reflects the displacement of the silicon atoms in the cubic to tetragonal phase transition.

References

5.2

Altmann, S. L. & Herzig, P. (1994). *Point-Group Theory Tables*. Oxford Science Publications.

International Tables for Crystallography (2004). Vol. C, edited by E. Prince, Table 8.3.1.1. Dordrecht: Kluwer Academic Publishers.

Megaw, H. D. (1973). *Crystal structures: a working approach*, pp. 259–262. Philadelphia: Saunders.

Nieuwenkamp, W. (1935). *Die Kristallstruktur des Tief-Cristobalits SiO₂. Z. Kristallogr.* **92**, 82–88.

Wyckoff, R. W. G. (1925). *Die Kristallstruktur von β-Cristobalit SiO₂ (bei hohen Temperaturen stabile Form). Z. Kristallogr.* **62**, 189–200.

6. THE 17 PLANE GROUPS (TWO-DIMENSIONAL SPACE GROUPS)

Diagrams of symmetry elements and of the general position

Origin

Asymmetric unit

Symmetry operations

Generators selected

Positions, with multiplicities, site symmetries, coordinates, reflection conditions

Maximal non-isomorphic subgroups

Maximal isomorphic subgroups of lowest index

Minimal non-isomorphic supergroups

No. 1   *p*1   Patterson symmetry *p*2

   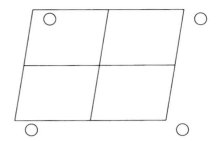

Origin arbitrary

Asymmetric unit $0 \le x \le 1$; $0 \le y \le 1$

Symmetry operations

(1) 1

Generators selected   (1); $t(1,0)$; $t(0,1)$

Positions

Multiplicity, Wyckoff letter, Site symmetry	Coordinates	Reflection conditions
		General:
1 *a* 1 (1) x, y		no conditions

Maximal non-isomorphic subgroups

I   none
IIa   none
IIb   none

Maximal isomorphic subgroups of lowest index

IIc   [2] *p*1 ($\mathbf{a}' = 2\mathbf{a}$ or $\mathbf{b}' = 2\mathbf{b}$ or $\mathbf{a}' = \mathbf{a} + \mathbf{b}, \mathbf{b}' = -\mathbf{a} + \mathbf{b}$) (1)

Minimal non-isomorphic supergroups

I   [2] *p*2 (2); [2] *pm* (3); [2] *pg* (4); [2] *cm* (5); [3] *p*3 (13)
II   none

Patterson symmetry *p*2   *p*2   No. 2

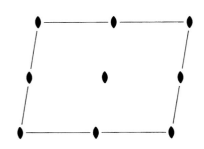 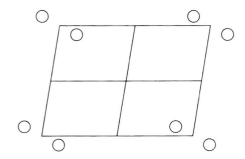

Origin at 2

Asymmetric unit $0 \le x \le \frac{1}{2}; \quad 0 \le y \le 1$

Symmetry operations

(1) 1   (2) 2   0,0

Generators selected   (1); $t(1,0)$; $t(0,1)$; (2)

Positions

Multiplicity, Wyckoff letter, Site symmetry		Coordinates	Reflection conditions
			General:
2	e	1 (1) x, y (2) $\bar{x}, \bar{y}$	no conditions
			Special: no extra conditions
1	d	2 $\frac{1}{2}, \frac{1}{2}$	
1	c	2 $\frac{1}{2}, 0$	
1	b	2 $0, \frac{1}{2}$	
1	a	2 $0, 0$	

Maximal non-isomorphic subgroups

I   [2] *p*1 (1)   1

IIa   none

IIb   none

Maximal isomorphic subgroups of lowest index

IIc   [2] *p*2 ($\mathbf{a}' = 2\mathbf{a}$ or $\mathbf{b}' = 2\mathbf{b}$ or $\mathbf{a}' = \mathbf{a} + \mathbf{b}, \mathbf{b}' = -\mathbf{a} + \mathbf{b}$) (2)

Minimal non-isomorphic supergroups

I   [2] *p*2*mm* (6); [2] *p*2*mg* (7); [2] *p*2*gg* (8); [2] *c*2*mm* (9); [2] *p*4 (10); [3] *p*6 (16)

II   none

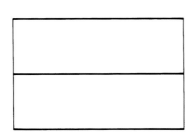
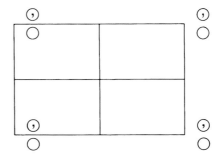

Origin on m

Asymmetric unit $0 \le x \le \frac{1}{2}$; $0 \le y \le 1$

Symmetry operations

(1) 1   (2) m $0, y$

Generators selected   (1); $t(1,0)$; $t(0,1)$; (2)

Positions

Multiplicity, Wyckoff letter, Site symmetry	Coordinates		Reflection conditions
			General:
2 c 1	(1) x, y (2) $\bar{x}, y$		no conditions
			Special: no extra conditions
1 b .m.	$\frac{1}{2}, y$		
1 a .m.	$0, y$		

Maximal non-isomorphic subgroups

I   [2] $p1$ (1)   1
IIa   none
IIb   [2] pg (**b**′ = 2**b**) (4); [2] cm (**a**′ = 2**a**, **b**′ = 2**b**) (5)

Maximal isomorphic subgroups of lowest index

IIc   [2] pm (**a**′ = 2**a**) (3); [2] pm (**b**′ = 2**b**) (3)

Minimal non-isomorphic supergroups

I   [2] $p2mm$ (6); [2] $p2mg$ (7)
II   [2] cm (5)

Rectangular m pg

Patterson symmetry $p2mm$ $p1g1$ **No. 4**

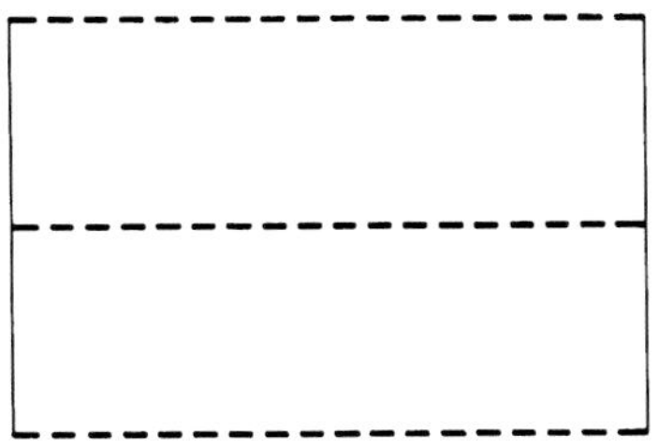

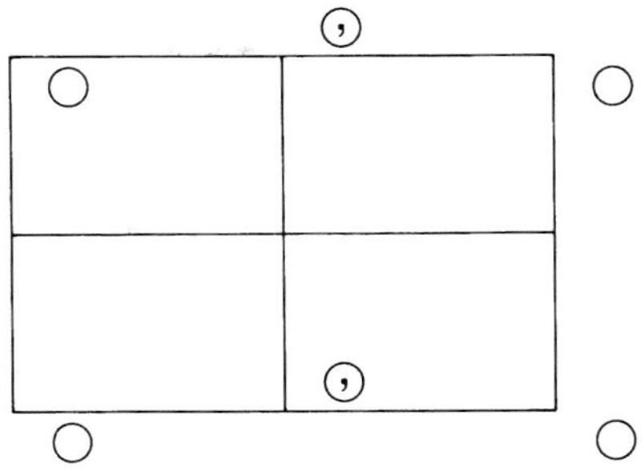

Origin on g

Asymmetric unit $0 \le x \le \frac{1}{2}$; $0 \le y \le 1$

Symmetry operations

(1) 1 (2) b $0,y$

Generators selected (1); $t(1,0)$; $t(0,1)$; (2)

Positions

Multiplicity, Wyckoff letter, Site symmetry	Coordinates		Reflection conditions
			General:
2 a 1	(1) x,y	(2) $\bar{x},y+\frac{1}{2}$	$0k$: $k=2n$

Maximal non-isomorphic subgroups

I [2] $p1$ (1) 1

IIa none

IIb none

Maximal isomorphic subgroups of lowest index

IIc [2] pg ($\mathbf{a}'=2\mathbf{a}$) (4); [3] pg ($\mathbf{b}'=3\mathbf{b}$) (4)

Minimal non-isomorphic supergroups

I [2] $p2mg$ (7); [2] $p2gg$ (8)

II [2] cm (5); [2] pm ($\mathbf{b}'=\frac{1}{2}\mathbf{b}$) (3)

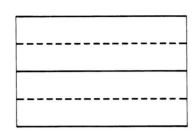 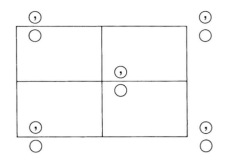

Origin on m

Asymmetric unit $0 \leq x \leq \frac{1}{2}$; $0 \leq y \leq \frac{1}{2}$

Symmetry operations

For $(0,0)+$ set

(1) 1   (2) m $0,y$

For $(\frac{1}{2},\frac{1}{2})+$ set

(1) $t(\frac{1}{2},\frac{1}{2})$ (2) b $\frac{1}{4},y$

Generators selected   (1); $t(1,0)$; $t(0,1)$; $t(\frac{1}{2},\frac{1}{2})$; (2)

Positions

Multiplicity, Wyckoff letter, Site symmetry	Coordinates $(0,0)+$ $(\frac{1}{2},\frac{1}{2})+$		Reflection conditions
4 b 1	(1) x,y	(2) $\bar{x},y$	General: hk: $h+k=2n$ $h0$: $h=2n$ $0k$: $k=2n$ Special: no extra conditions
2 a $.m.$	$0,y$		

Maximal non-isomorphic subgroups

I $[2]\,c\,1\,(p\,1,1)$ $1+$

IIa $[2]\,pg\,(4)$ $1;\ 2+(\frac{1}{2},\frac{1}{2})$

       $[2]\,pm\,(3)$ $1;\ 2$

IIb   none

Maximal isomorphic subgroups of lowest index

IIc $[3]\,cm\,(\mathbf{a}'=3\mathbf{a})\,(5)$; $[3]\,cm\,(\mathbf{b}'=3\mathbf{b})\,(5)$

Minimal non-isomorphic supergroups

I $[2]\,c\,2mm\,(9)$; $[3]\,p\,3m\,1\,(14)$; $[3]\,p\,3\,1\,m\,(15)$

II $[2]\,pm\,(\mathbf{a}'=\frac{1}{2}\mathbf{a},\mathbf{b}'=\frac{1}{2}\mathbf{b})\,(3)$

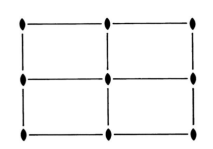 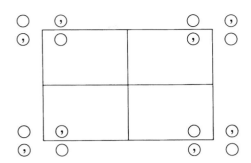

Origin at $2mm$

Asymmetric unit $0 \le x \le \frac{1}{2}$; $0 \le y \le \frac{1}{2}$

Symmetry operations

(1) 1   (2) 2 0,0   (3) m 0,y (4) m x,0

Generators selected   (1); $t(1,0)$; $t(0,1)$; (2); (3)

Positions

Multiplicity, Wyckoff letter, Site symmetry		Coordinates				Reflection conditions
						General:
4	i	1	(1) x,y (2) $\bar{x},\bar{y}$ (3) $\bar{x},y$ (4) $x,\bar{y}$			no conditions
						Special: no extra conditions
2	h	. m .	$\frac{1}{2},y$ $\frac{1}{2},\bar{y}$			
2	g	. m .	$0,y$ $0,\bar{y}$			
2	f	. . m	$x,\frac{1}{2}$ $\bar{x},\frac{1}{2}$			
2	e	. . m	$x,0$ $\bar{x},0$			
1	d	$2mm$	$\frac{1}{2},\frac{1}{2}$			
1	c	$2mm$	$\frac{1}{2},0$			
1	b	$2mm$	$0,\frac{1}{2}$			
1	a	$2mm$	$0,0$			

Maximal non-isomorphic subgroups

I   [2] $p1m1\,(pm,3)$ 1; 3
      [2] $p11m\,(pm,3)$ 1; 4
      [2] $p211\,(p2,2)$ 1; 2

IIa   none

IIb   [2] $p2mg\,(\mathbf{a}'=2\mathbf{a})\,(7)$; [2] $p2gm\,(\mathbf{b}'=2\mathbf{b})\,(p2mg,7)$; [2] $c2mm\,(\mathbf{a}'=2\mathbf{a},\mathbf{b}'=2\mathbf{b})\,(9)$

Maximal isomorphic subgroups of lowest index

IIc   [2] $p2mm\,(\mathbf{a}'=2\mathbf{a}$ or $\mathbf{b}'=2\mathbf{b})\,(6)$

Minimal non-isomorphic supergroups

I   [2] $p4mm\,(11)$
II   [2] $c2mm\,(9)$

$p2mg$ $2mm$ Rectangular

No. 7 $p2mg$

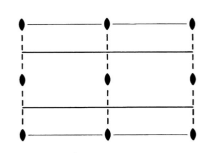 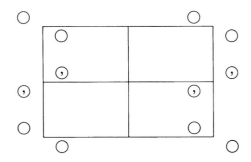

Origin at $2\,1\,g$

Asymmetric unit $0 \le x \le \frac{1}{4}; \quad 0 \le y \le 1$

Symmetry operations

(1) 1 (2) $2 \quad 0,0$ (3) $m \quad \frac{1}{4},y$ (4) $a \quad x,0$

Generators selected $(1); \ t(1,0); \ t(0,1); \ (2); \ (3)$

Positions

Multiplicity, Wyckoff letter, Site symmetry		Coordinates			Reflection conditions

General:

| 4 | d | 1 | (1) x,y | (2) $\bar{x},\bar{y}$ | (3) $\bar{x}+\frac{1}{2},y$ | (4) $x+\frac{1}{2},\bar{y}$ | $h0: \ h=2n$ |

Special: as above, plus

| 2 | c | $.m.$ | $\frac{1}{4},y$ | $\frac{3}{4},\bar{y}$ | no extra conditions |

| 2 | b | $2..$ | $0,\frac{1}{2}$ | $\frac{1}{2},\frac{1}{2}$ | $hk: \ h=2n$ |

| 2 | a | $2..$ | $0,0$ | $\frac{1}{2},0$ | $hk: \ h=2n$ |

Maximal non-isomorphic subgroups

I $[2]\ p11g\ (pg,4)$ $1;\ 4$
      $[2]\ p1m1\ (pm,3)$ $1;\ 3$
      $[2]\ p211\ (p2,2)$ $1;\ 2$

IIa   none

IIb $[2]\ p2gg\ (\mathbf{b}'=2\mathbf{b})\ (8)$

Maximal isomorphic subgroups of lowest index

IIc $[2]\ p2mg\ (\mathbf{b}'=2\mathbf{b})\ (7); \ [3]\ p2mg\ (\mathbf{a}'=3\mathbf{a})\ (7)$

Minimal non-isomorphic supergroups

I   none

II $[2]\ c2mm\ (9); \ [2]\ p2mm\ (\mathbf{a}'=\frac{1}{2}\mathbf{a})\ (6)$

Rectangular

$2mm$

$p2gg$

Patterson symmetry $p2mm$

$p2gg$

No. 8

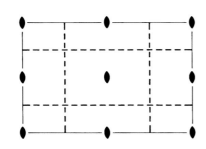
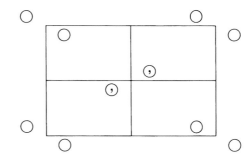

Origin at 2

Asymmetric unit $\quad 0 \le x \le \frac{1}{2}; \quad 0 \le y \le \frac{1}{2}$

Symmetry operations

(1) 1 (2) 2 0,0 (3) b $\frac{1}{4}, y$ (4) a $x, \frac{1}{4}$

Generators selected (1); $t(1,0)$; $t(0,1)$; (2); (3)

Positions

Multiplicity, Wyckoff letter, Site symmetry	Coordinates				Reflection conditions
					General:
4 c 1	(1) x,y	(2) $\bar{x},\bar{y}$	(3) $\bar{x}+\frac{1}{2}, y+\frac{1}{2}$	(4) $x+\frac{1}{2}, \bar{y}+\frac{1}{2}$	$h0$: $h=2n$
					$0k$: $k=2n$
					Special: as above, plus
2 b 2..	$\frac{1}{2},0$	$0,\frac{1}{2}$			hk: $h+k=2n$
2 a 2..	$0,0$	$\frac{1}{2},\frac{1}{2}$			hk: $h+k=2n$

Maximal non-isomorphic subgroups

I [2] $p1g1$ $(pg, 4)$ 1; 3
 [2] $p11g$ $(pg, 4)$ 1; 4
 [2] $p211$ $(p2, 2)$ 1; 2

IIa none

IIb none

Maximal isomorphic subgroups of lowest index

IIc [3] $p2gg$ ($\mathbf{a}' = 3\mathbf{a}$ or $\mathbf{b}' = 3\mathbf{b}$) (8)

Minimal non-isomorphic supergroups

I [2] $p4gm$ (12)

II [2] $c2mm$ (9); [2] $p2mg$ ($\mathbf{a}' = \frac{1}{2}\mathbf{a}$) (7); [2] $p2gm$ ($\mathbf{b}' = \frac{1}{2}\mathbf{b}$) ($p2mg$, 7)

$c\,2mm$ $\quad\quad 2mm$ $\quad\quad\quad\quad\quad$ Rectangular

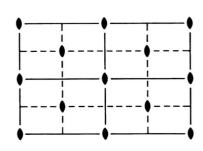 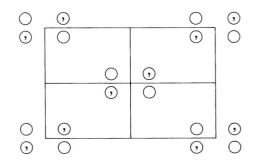

Origin at $2mm$

Asymmetric unit $\quad 0 \le x \le \frac{1}{4};\quad 0 \le y \le \frac{1}{2}$

Symmetry operations

For $(0,0)+$ set

(1) 1 $\quad\quad\quad$ (2) 2 $\quad 0,0$ $\quad\quad$ (3) m $\quad 0,y$ $\quad\quad$ (4) m $\quad x,0$

For $(\frac{1}{2},\frac{1}{2})+$ set

(1) $t(\frac{1}{2},\frac{1}{2})$ $\quad\quad$ (2) 2 $\quad \frac{1}{4},\frac{1}{4}$ $\quad\quad$ (3) b $\quad \frac{1}{4},y$ $\quad\quad$ (4) a $\quad x,\frac{1}{4}$

Generators selected $\quad$ (1); $t(1,0)$; $t(0,1)$; $t(\frac{1}{2},\frac{1}{2})$; (2); (3)

Positions

Multiplicity, Wyckoff letter, Site symmetry	Coordinates $(0,0)+ \quad (\frac{1}{2},\frac{1}{2})+$				Reflection conditions

General:

8	f	1	(1) x,y	(2) $\bar{x},\bar{y}$	(3) $\bar{x},y$	(4) $x,\bar{y}$

hk: $h+k = 2n$
$h0$: $h = 2n$
$0k$: $k = 2n$

Special: as above, plus

4	e	$.m.$	$0,y$	$0,\bar{y}$

no extra conditions

4	d	$..m$	$x,0$	$\bar{x},0$

no extra conditions

4	c	$2..$	$\frac{1}{4},\frac{1}{4}$	$\frac{3}{4},\frac{1}{4}$

hk: $h = 2n$

2	b	$2mm$	$0,\frac{1}{2}$

no extra conditions

2	a	$2mm$	$0,0$

no extra conditions

Maximal non-isomorphic subgroups

I $\quad$ [2] $c\,1m1\,(cm, 5)$ $\quad\quad$ (1; 3)+
$\quad\quad$ [2] $c\,11m\,(cm, 5)$ $\quad\quad$ (1; 4)+
$\quad\quad$ [2] $c\,211\,(p2, 2)$ $\quad\quad$ (1; 2)+

IIa $\quad$ [2] $p\,2gg\,(8)$ $\quad\quad\quad\quad$ 1; 2; (3; 4) $+ (\frac{1}{2},\frac{1}{2})$
$\quad\quad$ [2] $p\,2gm\,(p2mg, 7)$ $\quad$ 1; 4; (2; 3) $+ (\frac{1}{2},\frac{1}{2})$
$\quad\quad$ [2] $p\,2mg\,(7)$ $\quad\quad\quad$ 1; 3; (2; 4) $+ (\frac{1}{2},\frac{1}{2})$
$\quad\quad$ [2] $p\,2mm\,(6)$ $\quad\quad\quad$ 1; 2; 3; 4

IIb $\quad$ none

Maximal isomorphic subgroups of lowest index

IIc $\quad$ [3] $c\,2mm\,(\mathbf{a}' = 3\mathbf{a}$ or $\mathbf{b}' = 3\mathbf{b})\,(9)$

Minimal non-isomorphic supergroups

I $\quad$ [2] $p\,4mm\,(11)$; [2] $p\,4gm\,(12)$; [3] $p\,6mm\,(17)$
II $\quad$ [2] $p\,2mm\,(\mathbf{a}' = \frac{1}{2}\mathbf{a}, \mathbf{b}' = \frac{1}{2}\mathbf{b})\,(6)$

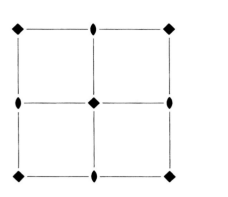
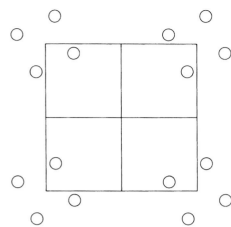

Origin at 4

Asymmetric unit $0 \leq x \leq \frac{1}{2}$; $0 \leq y \leq \frac{1}{2}$

Symmetry operations

(1) 1   (2) 2   0,0   (3) 4^+ 0,0   (4) 4^- 0,0

Generators selected   (1); $t(1,0)$; $t(0,1)$; (2); (3)

Positions

Multiplicity, Wyckoff letter, Site symmetry	Coordinates				Reflection conditions
					General:
4 *d* 1	(1) x,y	(2) $\bar{x},\bar{y}$	(3) $\bar{y},x$	(4) $y,\bar{x}$	no conditions
					Special:
2 *c* 2 . .	$\frac{1}{2},0$	$0,\frac{1}{2}$			hk: $h+k=2n$
1 *b* 4 . .	$\frac{1}{2},\frac{1}{2}$				no extra conditions
1 *a* 4 . .	0,0				no extra conditions

Maximal non-isomorphic subgroups

I   [2] *p* 2 (2)   1; 2
IIa   none
IIb   none

Maximal isomorphic subgroups of lowest index

IIc   [2] *c* 4 (**a**$'$ = 2**a**, **b**$'$ = 2**b**) (*p* 4, 10)

Minimal non-isomorphic supergroups

I   [2] *p* 4 *mm* (11); [2] *p* 4 *gm* (12)
II   none

p4mm   4mm   Square

No. 11   p4mm   Patterson symmetry p4mm

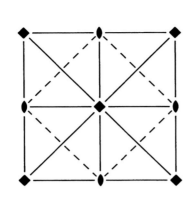
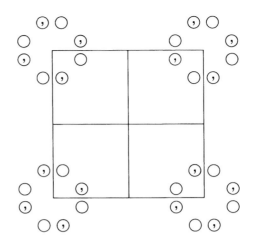

Origin at 4mm

Asymmetric unit $0 \leq x \leq \frac{1}{2}$; $0 \leq y \leq \frac{1}{2}$; $x \leq y$

Symmetry operations

(1) 1   (2) 2 0,0   (3) 4^+ 0,0   (4) 4^- 0,0
(5) m 0,y   (6) m x,0   (7) m x,x   (8) m x,$\bar{x}$

Generators selected   (1); $t(1,0)$; $t(0,1)$; (2); (3); (5)

Positions

Multiplicity, Wyckoff letter, Site symmetry	Coordinates				Reflection conditions

General:

8   g   1   (1) x,y (2) $\bar{x},\bar{y}$ (3) $\bar{y},x$ (4) $y,\bar{x}$ no conditions
(5) $\bar{x},y$ (6) $x,\bar{y}$ (7) y,x (8) $\bar{y},\bar{x}$

Special:

4   f   . . m x,x $\bar{x},\bar{x}$ $\bar{x},x$ $x,\bar{x}$ no extra conditions

4   e   . m . $x,\frac{1}{2}$ $\bar{x},\frac{1}{2}$ $\frac{1}{2},x$ $\frac{1}{2},\bar{x}$ no extra conditions

4   d   . m . $x,0$ $\bar{x},0$ $0,x$ $0,\bar{x}$ no extra conditions

2   c   2 m m . $\frac{1}{2},0$ $0,\frac{1}{2}$ hk: $h+k=2n$

1   b   4 m m $\frac{1}{2},\frac{1}{2}$ no extra conditions

1   a   4 m m $0,0$ no extra conditions

Maximal non-isomorphic subgroups

I   [2] $p411\,(p4,\,10)$ 1; 2; 3; 4
    [2] $p21m\,(c2mm,\,9)$ 1; 2; 7; 8
    [2] $p2m1\,(p2mm,\,6)$ 1; 2; 5; 6

IIa   none

IIb   [2] $c4mg\,(\mathbf{a}'=2\mathbf{a},\mathbf{b}'=2\mathbf{b})\,(p4gm,\,12)$

Maximal isomorphic subgroups of lowest index

IIc   [2] $c4mm\,(\mathbf{a}'=2\mathbf{a},\mathbf{b}'=2\mathbf{b})\,(p4mm,\,11)$

Minimal non-isomorphic supergroups

I   none

II   none

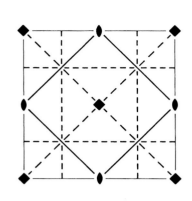
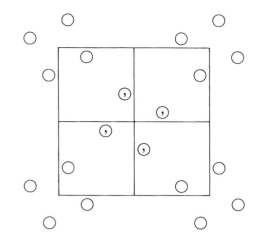

Origin at $41g$

Asymmetric unit $0 \le x \le \frac{1}{2}$; $0 \le y \le \frac{1}{2}$; $y \le \frac{1}{2} - x$

Symmetry operations

(1) 1   (2) 2 0,0   (3) 4^+ 0,0   (4) 4^- 0,0
(5) b $\frac{1}{4}, y$ (6) a $x, \frac{1}{4}$ (7) $g(\frac{1}{2}, \frac{1}{2})$ x, x (8) m $x + \frac{1}{2}, \bar{x}$

Generators selected   (1); $t(1,0)$; $t(0,1)$; (2); (3); (5)

Positions

Multiplicity, Wyckoff letter, Site symmetry	Coordinates				Reflection conditions
					General:
8 d 1	(1) x,y (2) $\bar{x}, \bar{y}$ (3) $\bar{y}, x$ (4) $y, \bar{x}$				$h0$: $h = 2n$
	(5) $\bar{x} + \frac{1}{2}, y + \frac{1}{2}$ (6) $x + \frac{1}{2}, \bar{y} + \frac{1}{2}$ (7) $y + \frac{1}{2}, x + \frac{1}{2}$ (8) $\bar{y} + \frac{1}{2}, \bar{x} + \frac{1}{2}$				$0k$: $k = 2n$
					Special: as above, plus
4 c $..m$	$x, x + \frac{1}{2}$ $\bar{x}, \bar{x} + \frac{1}{2}$ $\bar{x} + \frac{1}{2}, x$ $x + \frac{1}{2}, \bar{x}$				no extra conditions
2 b $2.mm$	$\frac{1}{2}, 0$ $0, \frac{1}{2}$				hk: $h + k = 2n$
2 a $4..$	$0,0$ $\frac{1}{2}, \frac{1}{2}$				hk: $h + k = 2n$

Maximal non-isomorphic subgroups

I   [2] $p411$ ($p4$, 10)   1; 2; 3; 4
        [2] $p21m$ ($c2mm$, 9)   1; 2; 7; 8
        [2] $p2g1$ ($p2gg$, 8)   1; 2; 5; 6
IIa   none
IIb   none

Maximal isomorphic subgroups of lowest index

IIc   [9] $p4gm$ ($\mathbf{a}' = 3\mathbf{a}, \mathbf{b}' = 3\mathbf{b}$) (12)

Minimal non-isomorphic supergroups

I   none
II   [2] $c4gm$ ($p4mm$, 11)

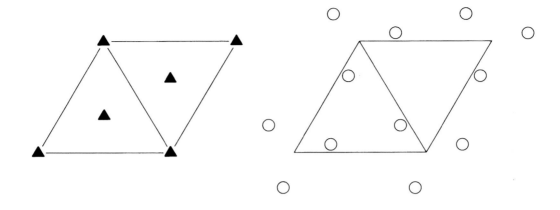

Origin at 3

Asymmetric unit $0 \leq x \leq \frac{2}{3}$; $0 \leq y \leq \frac{2}{3}$; $x \leq (1+y)/2$; $y \leq \min(1-x, (1+x)/2)$

Vertices $0,0$ $\frac{1}{2},0$ $\frac{2}{3},\frac{1}{3}$ $\frac{1}{3},\frac{2}{3}$ $0,\frac{1}{2}$

Symmetry operations

(1) 1   (2) 3^+ 0,0   (3) 3^- 0,0

Generators selected   (1); $t(1,0)$; $t(0,1)$; (2)

Positions

Multiplicity,   Coordinates   Reflection conditions
Wyckoff letter,
Site symmetry

General:

3   *d*   1   (1) x,y (2) $\bar{y}, x-y$ (3) $\bar{x}+y, \bar{x}$ no conditions

Special: no extra conditions

1   *c*   3 . . $\frac{2}{3}, \frac{1}{3}$

1   *b*   3 . . $\frac{1}{3}, \frac{2}{3}$

1   *a*   3 . . $0,0$

Maximal non-isomorphic subgroups

I   [3] *p*1 (1)   1
IIa   none
IIb   none

Maximal isomorphic subgroups of lowest index

IIc   [3] *h*3 ($\mathbf{a}' = 3\mathbf{a}, \mathbf{b}' = 3\mathbf{b}$) (*p*3, 13)

Minimal non-isomorphic supergroups

I   [2] *p*3*m*1 (14); [2] *p*31*m* (15); [2] *p*6 (16)
II   none

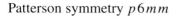

Patterson symmetry $p6mm$ $p3m1$ No. 14

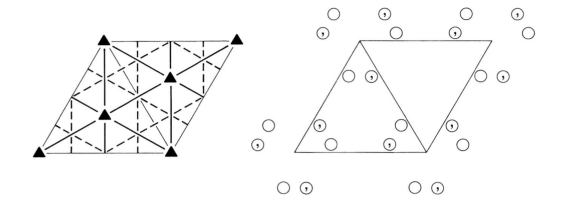

Origin at $3m1$

Asymmetric unit $0 \le x \le \frac{2}{3};$ $0 \le y \le \frac{2}{3};$ $x \le 2y;$ $y \le \min(1-x, 2x)$

         Vertices $0,0$ $\frac{2}{3}, \frac{1}{3}$ $\frac{1}{3}, \frac{2}{3}$

Symmetry operations

(1) 1 (2) 3^+ $0,0$ (3) 3^- $0,0$
(4) m $x,\bar{x}$ (5) m $x,2x$ (6) m $2x,x$

Generators selected $(1);$ $t(1,0);$ $t(0,1);$ $(2);$ (4)

Positions

Multiplicity, Wyckoff letter, Site symmetry		Coordinates		Reflection conditions
				General:
6 e 1	(1) x,y (2) $\bar{y}, x-y$ (3) $\bar{x}+y, \bar{x}$			no conditions
	(4) $\bar{y}, \bar{x}$ (5) $\bar{x}+y, y$ (6) $x, x-y$			
				Special: no extra conditions
3 d .m.	$x,\bar{x}$ $x,2x$ $2\bar{x},\bar{x}$			
1 c $3m$.	$\frac{2}{3}, \frac{1}{3}$			
1 b $3m$.	$\frac{1}{3}, \frac{2}{3}$			
1 a $3m$.	$0,0$			

Maximal non-isomorphic subgroups

I   [2] $p311$ ($p3$, 13)   1; 2; 3
      $\left\{ \begin{array}{l} \text{[3] } p1m1 \, (cm, 5) \quad 1; 4 \\ \text{[3] } p1m1 \, (cm, 5) \quad 1; 5 \\ \text{[3] } p1m1 \, (cm, 5) \quad 1; 6 \end{array} \right.$

IIa   none
IIb   [3] $h3m1$ ($\mathbf{a}' = 3\mathbf{a}, \mathbf{b}' = 3\mathbf{b}$) ($p31m$, 15)

Maximal isomorphic subgroups of lowest index
IIc   [4] $p3m1$ ($\mathbf{a}' = 2\mathbf{a}, \mathbf{b}' = 2\mathbf{b}$) (14)

Minimal non-isomorphic supergroups
I   [2] $p6mm$ (17)
II   [3] $h3m1$ ($p31m$, 15)

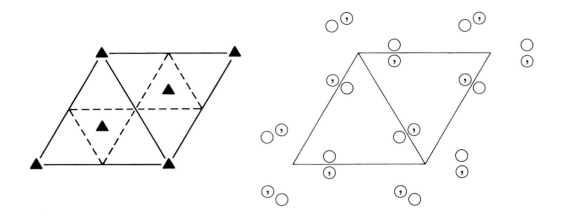

Origin at $31m$

Asymmetric unit $0 \le x \le \frac{2}{3};$ $0 \le y \le \frac{1}{2};$ $x \le (1+y)/2;$ $y \le \min(1-x,x)$

         Vertices $0,0$ $\frac{1}{2},0$ $\frac{2}{3},\frac{1}{3}$ $\frac{1}{2},\frac{1}{2}$

Symmetry operations

(1) 1   (2) 3^+ $0,0$ (3) 3^- $0,0$
(4) m x,x (5) m $x,0$ (6) m $0,y$

Generators selected   (1); $t(1,0)$; $t(0,1)$; (2); (4)

Positions

Multiplicity,   Coordinates   Reflection conditions
Wyckoff letter,
Site symmetry   General:

6   d 1   (1) x,y (2) $\bar{y},x-y$ (3) $\bar{x}+y,\bar{x}$ no conditions
                   (4) y,x (5) $x-y,\bar{y}$ (6) $\bar{x},\bar{x}+y$

                                                       Special: no extra conditions

3   c $. . m$ $x,0$ $0,x$ $\bar{x},\bar{x}$

2   b $3 . .$ $\frac{1}{3},\frac{2}{3}$ $\frac{2}{3},\frac{1}{3}$

1   a $3 . m$ $0,0$

Maximal non-isomorphic subgroups

I   [2] $p311$ $(p3, 13)$ 1; 2; 3
        ⎧ [3] $p11m$ $(cm, 5)$ 1; 4
        ⎨ [3] $p11m$ $(cm, 5)$ 1; 5
        ⎩ [3] $p11m$ $(cm, 5)$ 1; 6

IIa   none
IIb   [3] $h31m$ ($\mathbf{a}' = 3\mathbf{a}, \mathbf{b}' = 3\mathbf{b}$) $(p3m1, 14)$

Maximal isomorphic subgroups of lowest index

IIc   [4] $p31m$ ($\mathbf{a}' = 2\mathbf{a}, \mathbf{b}' = 2\mathbf{b}$) (15)

Minimal non-isomorphic supergroups

I   [2] $p6mm$ (17)
II   [3] $h31m$ $(p3m1, 14)$

Generators selected   (1); $t(1,0,0)$; $t(0,1,0)$; $t(0,0,1)$; (2)

Positions

Multiplicity, Wyckoff letter, Site symmetry	Coordinates	Reflection conditions
		General:
2 i 1	(1) x,y,z (2) $\bar{x},\bar{y},\bar{z}$	no conditions
		Special: no extra conditions

1	h	$\bar{1}$	$\frac{1}{2},\frac{1}{2},\frac{1}{2}$
1	g	$\bar{1}$	$0,\frac{1}{2},\frac{1}{2}$
1	f	$\bar{1}$	$\frac{1}{2},0,\frac{1}{2}$
1	e	$\bar{1}$	$\frac{1}{2},\frac{1}{2},0$
1	d	$\bar{1}$	$\frac{1}{2},0,0$
1	c	$\bar{1}$	$0,\frac{1}{2},0$
1	b	$\bar{1}$	$0,0,\frac{1}{2}$
1	a	$\bar{1}$	$0,0,0$

Symmetry of special projections

Along $[001]$ $p2$	Along $[100]$ $p2$	Along $[010]$ $p2$
$\mathbf{a}' = \mathbf{a}_p$ $\mathbf{b}' = \mathbf{b}_p$	$\mathbf{a}' = \mathbf{b}_p$ $\mathbf{b}' = \mathbf{c}_p$	$\mathbf{a}' = \mathbf{c}_p$ $\mathbf{b}' = \mathbf{a}_p$
Origin at $0,0,z$	Origin at $x,0,0$	Origin at $0,y,0$

Maximal non-isomorphic subgroups

I   [2] $P1$ (1)   1

IIa   none

IIb   none

Maximal isomorphic subgroups of lowest index

IIc   [2] $P\bar{1}$ ($\mathbf{a}' = 2\mathbf{a}$ or $\mathbf{b}' = 2\mathbf{b}$ or $\mathbf{c}' = 2\mathbf{c}$ or $\mathbf{b}' = \mathbf{b}+\mathbf{c}, \mathbf{c}' = -\mathbf{b}+\mathbf{c}$ or $\mathbf{a}' = \mathbf{a}-\mathbf{c}, \mathbf{c}' = \mathbf{a}+\mathbf{c}$ or $\mathbf{a}' = \mathbf{a}+\mathbf{b}, \mathbf{b}' = -\mathbf{a}+\mathbf{b}$ or $\mathbf{a}' = \mathbf{b}+\mathbf{c}, \mathbf{b}' = \mathbf{a}+\mathbf{c}, \mathbf{c}' = \mathbf{a}+\mathbf{b}$) (2)

Minimal non-isomorphic supergroups

I   [2] $P2/m$ (10); [2] $P2_1/m$ (11); [2] $C2/m$ (12); [2] $P2/c$ (13); [2] $P2_1/c$ (14); [2] $C2/c$ (15); [3] $P\bar{3}$ (147); [3] $R\bar{3}$ (148)

II   none

UNIQUE AXIS b

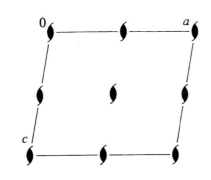

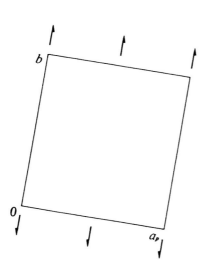
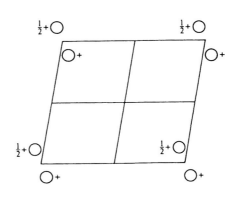

Origin on 2_1

Asymmetric unit $0 \le x \le 1; \quad 0 \le y \le 1; \quad 0 \le z \le \frac{1}{2}$

Symmetry operations

(1) 1   (2) $2(0,\frac{1}{2},0)$ $0,y,0$

Generators selected   (1); $t(1,0,0)$; $t(0,1,0)$; $t(0,0,1)$; (2)

Positions

Multiplicity, Wyckoff letter, Site symmetry	Coordinates	Reflection conditions
		General:
2 a 1	(1) x,y,z (2) $\bar{x}, y+\frac{1}{2}, \bar{z}$	$0k0: \ k=2n$

Symmetry of special projections

Along [001] $p1g1$
$\mathbf{a}' = \mathbf{a}_p$ $\mathbf{b}' = \mathbf{b}$
Origin at $0,0,z$

Along [100] $p11g$
$\mathbf{a}' = \mathbf{b}$ $\mathbf{b}' = \mathbf{c}_p$
Origin at $x,0,0$

Along [010] $p2$
$\mathbf{a}' = \mathbf{c}$ $\mathbf{b}' = \mathbf{a}$
Origin at $0,y,0$

Maximal non-isomorphic subgroups

I   [2] $P1$ (1)   1

IIa   none

IIb   none

Maximal isomorphic subgroups of lowest index

IIc   [2] $P12_1 1$ ($\mathbf{c}' = 2\mathbf{c}$ or $\mathbf{a}' = 2\mathbf{a}$ or $\mathbf{a}' = \mathbf{a}+\mathbf{c}, \mathbf{c}' = -\mathbf{a}+\mathbf{c}$) ($P2_1$, 4); [3] $P12_1 1$ ($\mathbf{b}' = 3\mathbf{b}$) ($P2_1$, 4)

Minimal non-isomorphic supergroups

I   [2] $P2_1/m$ (11); [2] $P2_1/c$ (14); [2] $P222_1$ (17); [2] $P2_12_12$ (18); [2] $P2_12_12_1$ (19); [2] $C222_1$ (20); [2] $Pmc2_1$ (26); [2] $Pca2_1$ (29); [2] $Pmn2_1$ (31); [2] $Pna2_1$ (33); [2] $Cmc2_1$ (36); [2] $P4_1$ (76); [2] $P4_3$ (78); [3] $P6_1$ (169); [3] $P6_5$ (170); [3] $P6_3$ (173)

II   [2] $C121$ ($C2$, 5); [2] $A121$ ($C2$, 5); [2] $I121$ ($C2$, 5); [2] $P121$ ($\mathbf{b}' = \frac{1}{2}\mathbf{b}$) ($P2$, 3)

UNIQUE AXIS b, CELL CHOICE 1

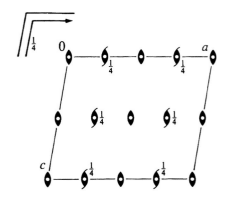

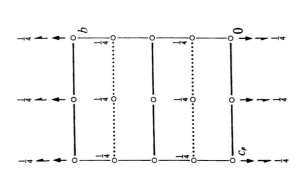

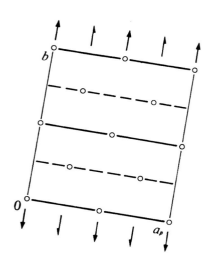

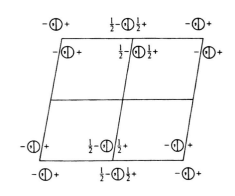

Origin at centre $(2/m)$

Asymmetric unit $0 \leq x \leq \frac{1}{2}$; $0 \leq y \leq \frac{1}{4}$; $0 \leq z \leq 1$

Symmetry operations

For $(0,0,0)+$ set

(1) 1   (2) 2 $0,y,0$ (3) $\bar{1}$ $0,0,0$ (4) m $x,0,z$

For $(\frac{1}{2},\frac{1}{2},0)+$ set

(1) $t(\frac{1}{2},\frac{1}{2},0)$ (2) $2(0,\frac{1}{2},0)$ $\frac{1}{4},y,0$ (3) $\bar{1}$ $\frac{1}{4},\frac{1}{4},0$ (4) a $x,\frac{1}{4},z$

Generators selected   (1); $t(1,0,0)$; $t(0,1,0)$; $t(0,0,1)$; $t(\frac{1}{2},\frac{1}{2},0)$; (2); (3)

Positions

Multiplicity, Wyckoff letter, Site symmetry		Coordinates $(0,0,0)+$ $(\frac{1}{2},\frac{1}{2},0)+$				Reflection conditions

General:

8	j	1	(1) x,y,z	(2) $\bar{x},y,\bar{z}$	(3) $\bar{x},\bar{y},\bar{z}$	(4) $x,\bar{y},z$

$hkl : h+k = 2n$
$h0l : h = 2n$
$0kl : k = 2n$
$hk0 : h+k = 2n$
$0k0 : k = 2n$
$h00 : h = 2n$

Special: as above, plus

4	i	m	$x,0,z$	$\bar{x},0,\bar{z}$	no extra conditions
4	h	2	$0,y,\frac{1}{2}$	$0,\bar{y},\frac{1}{2}$	no extra conditions
4	g	2	$0,y,0$	$0,\bar{y},0$	no extra conditions
4	f	$\bar{1}$	$\frac{1}{4},\frac{1}{4},\frac{1}{2}$	$\frac{3}{4},\frac{1}{4},\frac{1}{2}$	$hkl : h = 2n$
4	e	$\bar{1}$	$\frac{1}{4},\frac{1}{4},0$	$\frac{3}{4},\frac{1}{4},0$	$hkl : h = 2n$
2	d	$2/m$	$0,\frac{1}{2},\frac{1}{2}$		no extra conditions
2	c	$2/m$	$0,0,\frac{1}{2}$		no extra conditions
2	b	$2/m$	$0,\frac{1}{2},0$		no extra conditions
2	a	$2/m$	$0,0,0$		no extra conditions

Symmetry of special projections

Along [001] $c2mm$
$\mathbf{a}' = \mathbf{a}_p$ $\mathbf{b}' = \mathbf{b}$
Origin at $0,0,z$

Along [100] $p2mm$
$\mathbf{a}' = \frac{1}{2}\mathbf{b}$ $\mathbf{b}' = \mathbf{c}_p$
Origin at $x,0,0$

Along [010] $p2$
$\mathbf{a}' = \mathbf{c}$ $\mathbf{b}' = \frac{1}{2}\mathbf{a}$
Origin at $0,y,0$

Maximal non-isomorphic subgroups

I   [2] $C1m1$ $(Cm, 8)$ (1; 4)+   m
     [2] $C121$ $(C2, 5)$ (1; 2)+   2
     [2] $C\bar{1}$ $(P\bar{1}, 2)$ (1; 3)+ $\bar{1}$

IIa   [2] $P12_1/a1$ $(P2_1/c, 14)$ 1; 3; (2; 4) + $(\frac{1}{2},\frac{1}{2},0)$
     [2] $P12/a1$ $(P2/c, 13)$ 1; 2; (3; 4) + $(\frac{1}{2},\frac{1}{2},0)$
     [2] $P12_1/m1$ $(P2_1/m, 11)$ 1; 4; (2; 3) + $(\frac{1}{2},\frac{1}{2},0)$
     [2] $P12/m1$ $(P2/m, 10)$ 1; 2; 3; 4

IIb   [2] $C12/c1$ $(\mathbf{c}' = 2\mathbf{c})$ $(C2/c, 15)$; [2] $I12/c1$ $(\mathbf{c}' = 2\mathbf{c})$ $(C2/c, 15)$

Maximal isomorphic subgroups of lowest index

IIc   [2] $C12/m1$ $(\mathbf{c}' = 2\mathbf{c}$ or $\mathbf{a}' = \mathbf{a} + 2\mathbf{c}, \mathbf{c}' = 2\mathbf{c})$ $(C2/m, 12)$; [3] $C12/m1$ $(\mathbf{b}' = 3\mathbf{b})$ $(C2/m, 12)$

Minimal non-isomorphic supergroups

I   [2] $Cmcm$ (63); [2] $Cmce$ (64); [2] $Cmmm$ (65); [2] $Cmme$ (67); [2] $Fmmm$ (69); [2] $Immm$ (71); [2] $Ibam$ (72); [2] $Imma$ (74); [2] $I4/m$ (87); [3] $P\bar{3}1m$ (162); [3] $P\bar{3}m1$ (164); [3] $R\bar{3}m$ (166)

II   [2] $P12/m1$ $(\mathbf{a}' = \frac{1}{2}\mathbf{a}, \mathbf{b}' = \frac{1}{2}\mathbf{b})$ $(P2/m, 10)$

Cmcm#63: mmm
Cmce#64: mmm? ←
Cmmm#65: mmm
Cmme #67: mmm? ←
Fmmm#69: mmm
Immm#71: mmm

Ibam#72: mmm?
Imma#74: mmm
I4/m#87: 4/m
P$\bar{3}$1m#162: $\bar{3}$m
P$\bar{3}$m1 #164: $\bar{3}$m

R$\bar{3}$m#166: $\bar{3}$m

UNIQUE AXIS b, DIFFERENT CELL CHOICES

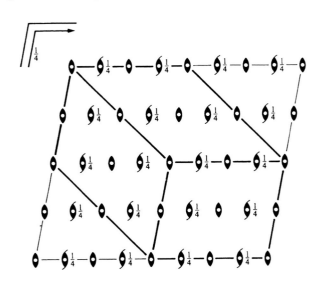
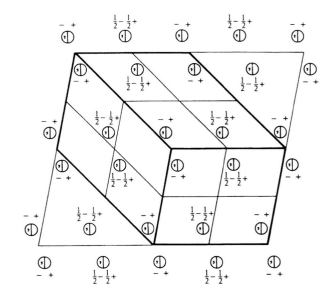

$C12/m1$

UNIQUE AXIS b, CELL CHOICE 1

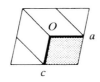

Origin at centre $(2/m)$

Asymmetric unit $0 \leq x \leq \frac{1}{2}$; $0 \leq y \leq \frac{1}{4}$; $0 \leq z \leq 1$

Generators selected   (1); $t(1,0,0)$; $t(0,1,0)$; $t(0,0,1)$; $t(\frac{1}{2},\frac{1}{2},0)$; (2); (3)

Positions

Multiplicity,
Wyckoff letter,
Site symmetry

			Coordinates				Reflection conditions

$(0,0,0)+$ $(\frac{1}{2},\frac{1}{2},0)+$

General:

8	j	1	(1) x,y,z	(2) $\bar{x},y,\bar{z}$	(3) $\bar{x},\bar{y},\bar{z}$	(4) $x,\bar{y},z$

hkl : $h+k=2n$ $hk0$: $h+k=2n$
$h0l$: $h=2n$ $0k0$: $k=2n$
$0kl$: $k=2n$ $h00$: $h=2n$

Special: as above, plus

4	i	m	$x,0,z$	$\bar{x},0,\bar{z}$					no extra conditions
4	h	2	$0,y,\frac{1}{2}$	$0,\bar{y},\frac{1}{2}$	4	g	2	$0,y,0$ $0,\bar{y},0$	no extra conditions
4	f	$\bar{1}$	$\frac{1}{4},\frac{1}{4},\frac{1}{2}$	$\frac{3}{4},\frac{1}{4},\frac{1}{2}$	4	e	$\bar{1}$	$\frac{1}{4},\frac{1}{4},0$ $\frac{3}{4},\frac{1}{4},0$	hkl : $h=2n$
2	d	$2/m$	$0,\frac{1}{2},\frac{1}{2}$		2	c	$2/m$	$0,0,\frac{1}{2}$	no extra conditions
2	b	$2/m$	$0,\frac{1}{2},0$		2	a	$2/m$	$0,0,0$	no extra conditions

Generators selected   (1); $t(1,0,0)$; $t(0,1,0)$; $t(0,0,1)$; (2); (3)

Positions

Multiplicity, Wyckoff letter, Site symmetry		Coordinates				Reflection conditions
						General:
4	c	1	(1) x,y,z (2) $\bar{x},\bar{y},z$ (3) $\bar{x}+\frac{1}{2},y+\frac{1}{2},\bar{z}$ (4) $x+\frac{1}{2},\bar{y}+\frac{1}{2},\bar{z}$				$h00:\ h=2n$
						$0k0:\ k=2n$
						Special: as above, plus
2	b	$..2$	$0,\frac{1}{2},z$ $\frac{1}{2},0,\bar{z}$			$hk0:\ h+k=2n$
2	a	$..2$	$0,0,z$ $\frac{1}{2},\frac{1}{2},\bar{z}$			$hk0:\ h+k=2n$

Symmetry of special projections

Along [001] $p2gg$
$\mathbf{a}'=\mathbf{a}$ $\mathbf{b}'=\mathbf{b}$
Origin at $0,0,z$

Along [100] $p2mg$
$\mathbf{a}'=\mathbf{b}$ $\mathbf{b}'=\mathbf{c}$
Origin at $x,\frac{1}{4},0$

Along [010] $p2gm$
$\mathbf{a}'=\mathbf{c}$ $\mathbf{b}'=\mathbf{a}$
Origin at $\frac{1}{4},y,0$

Maximal non-isomorphic subgroups

I $[2]\,P12_11\,(P2_1,4)$ 1; 3
      $[2]\,P2_111\,(P2_1,4)$ 1; 4
      $[2]\,P112\,(P2,3)$ 1; 2

IIa   none
IIb $[2]\,P2_12_12_1\,(\mathbf{c}'=2\mathbf{c})\,(19)$

Maximal isomorphic subgroups of lowest index

IIc $[2]\,P2_12_12\,(\mathbf{c}'=2\mathbf{c})\,(18)$; $[3]\,P2_12_12\,(\mathbf{a}'=3\mathbf{a}\text{ or }\mathbf{b}'=3\mathbf{b})\,(18)$

Minimal non-isomorphic supergroups

I $[2]\,Pbam\,(55)$; $[2]\,Pccn\,(56)$; $[2]\,Pbcm\,(57)$; $[2]\,Pnnm\,(58)$; $[2]\,Pmmn\,(59)$; $[2]\,Pbcn\,(60)$; $[2]\,P42_12\,(90)$; $[2]\,P4_22_12\,(94)$; $[2]\,P\bar{4}2_1m\,(113)$; $[2]\,P\bar{4}2_1c\,(114)$

II $[2]\,A2_122\,(C222_1,20)$; $[2]\,B22_12\,(C222_1,20)$; $[2]\,C222\,(21)$; $[2]\,I222\,(23)$; $[2]\,P22_12\,(\mathbf{a}'=\frac{1}{2}\mathbf{a})\,(P222_1,17)$; $[2]\,P2_122\,(\mathbf{b}'=\frac{1}{2}\mathbf{b})\,(P222_1,17)$

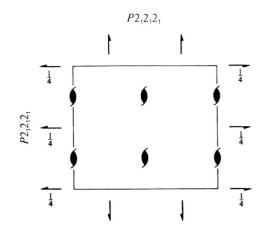

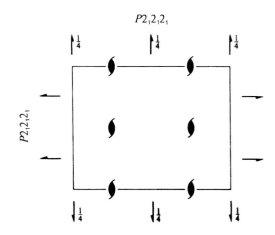

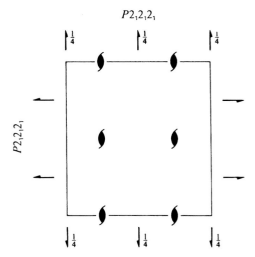

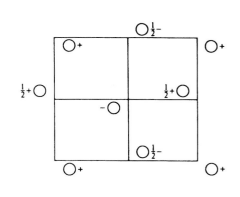

Origin at midpoint of three non-intersecting pairs of parallel 2_1 axes

Asymmetric unit $0 \le x \le \frac{1}{2}$; $0 \le y \le \frac{1}{2}$; $0 \le z \le 1$

Symmetry operations

(1) 1   (2) $2(0,0,\frac{1}{2})$ $\frac{1}{4},0,z$ (3) $2(0,\frac{1}{2},0)$ $0,y,\frac{1}{4}$ (4) $2(\frac{1}{2},0,0)$ $x,\frac{1}{4},0$

Generators selected   (1); $t(1,0,0)$; $t(0,1,0)$; $t(0,0,1)$; (2); (3)

Positions

Multiplicity, Wyckoff letter, Site symmetry	Coordinates				Reflection conditions
					General:
4 a 1	(1) x,y,z	(2) $\bar{x}+\frac{1}{2},\bar{y},z+\frac{1}{2}$	(3) $\bar{x},y+\frac{1}{2},\bar{z}+\frac{1}{2}$	(4) $x+\frac{1}{2},\bar{y}+\frac{1}{2},\bar{z}$	$h00:\ h=2n$
					$0k0:\ k=2n$
					$00l:\ l=2n$

Symmetry of special projections

Along $[001]$ $p2gg$
$\mathbf{a}'=\mathbf{a}$ $\mathbf{b}'=\mathbf{b}$
Origin at $\frac{1}{4},0,z$

Along $[100]$ $p2gg$
$\mathbf{a}'=\mathbf{b}$ $\mathbf{b}'=\mathbf{c}$
Origin at $x,\frac{1}{4},0$

Along $[010]$ $p2gg$
$\mathbf{a}'=\mathbf{c}$ $\mathbf{b}'=\mathbf{a}$
Origin at $0,y,\frac{1}{4}$

Maximal non-isomorphic subgroups

I   [2] $P112_1$ $(P2_1,4)$ 1; 2
      [2] $P12_11$ $(P2_1,4)$ 1; 3
      [2] $P2_111$ $(P2_1,4)$ 1; 4

IIa   none

IIb   none

Maximal isomorphic subgroups of lowest index

IIc   [3] $P2_12_12_1$ $(\mathbf{a}'=3\mathbf{a}$ or $\mathbf{b}'=3\mathbf{b}$ or $\mathbf{c}'=3\mathbf{c})$ (19)

Minimal non-isomorphic supergroups

I   [2] $Pbca$ (61); [2] $Pnma$ (62); [2] $P4_12_12$ (92); [2] $P4_32_12$ (96); [3] $P2_13$ (198)

II   [2] $A2_122$ $(C222_1,20)$; [2] $B22_12$ $(C222_1,20)$; [2] $C222_1$ (20); [2] $I2_12_12_1$ (24); [2] $P22_12_1$ $(\mathbf{a}'=\frac{1}{2}\mathbf{a})$ $(P2_12_12,18)$;
     [2] $P2_122_1$ $(\mathbf{b}'=\frac{1}{2}\mathbf{b})$ $(P2_12_12,18)$; [2] $P2_12_12$ $(\mathbf{c}'=\frac{1}{2}\mathbf{c})$ (18)

$Fdd2$ C_{2v}^{19} $mm2$ Orthorhombic

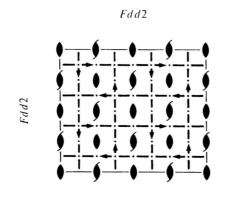

Fdd2

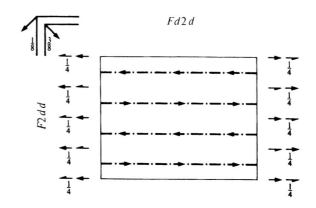

Fd2d

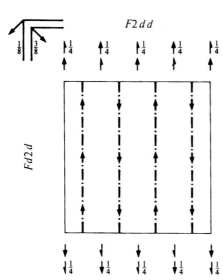

F2dd

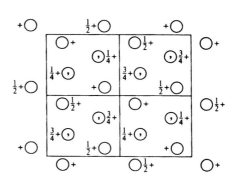

Origin on 1 1 2

Asymmetric unit $0 \le x \le \frac{1}{4}$; $0 \le y \le \frac{1}{4}$; $0 \le z \le 1$

Symmetry operations

For $(0,0,0)+$ set

(1) 1 (2) 2 $0,0,z$ (3) $d(\frac{1}{4},0,\frac{1}{4})$ $x,\frac{1}{8},z$ (4) $d(0,\frac{1}{4},\frac{1}{4})$ $\frac{1}{8},y,z$

For $(0,\frac{1}{2},\frac{1}{2})+$ set

(1) $t(0,\frac{1}{2},\frac{1}{2})$ (2) $2(0,0,\frac{1}{2})$ $0,\frac{1}{4},z$ (3) $d(\frac{1}{4},0,\frac{3}{4})$ $x,\frac{3}{8},z$ (4) $d(0,\frac{3}{4},\frac{3}{4})$ $\frac{1}{8},y,z$

For $(\frac{1}{2},0,\frac{1}{2})+$ set

(1) $t(\frac{1}{2},0,\frac{1}{2})$ (2) $2(0,0,\frac{1}{2})$ $\frac{1}{4},0,z$ (3) $d(\frac{3}{4},0,\frac{3}{4})$ $x,\frac{1}{8},z$ (4) $d(0,\frac{1}{4},\frac{3}{4})$ $\frac{3}{8},y,z$

For $(\frac{1}{2},\frac{1}{2},0)+$ set

(1) $t(\frac{1}{2},\frac{1}{2},0)$ (2) 2 $\frac{1}{4},\frac{1}{4},z$ (3) $d(\frac{3}{4},0,\frac{1}{4})$ $x,\frac{3}{8},z$ (4) $d(0,\frac{3}{4},\frac{1}{4})$ $\frac{3}{8},y,z$

Generators selected   (1); $t(1,0,0)$; $t(0,1,0)$; $t(0,0,1)$; $t(0,\frac{1}{2},\frac{1}{2})$; $t(\frac{1}{2},0,\frac{1}{2})$; (2); (3)

Positions

Multiplicity, Wyckoff letter, Site symmetry	Coordinates $(0,0,0)+$ $(0,\frac{1}{2},\frac{1}{2})+$ $(\frac{1}{2},0,\frac{1}{2})+$ $(\frac{1}{2},\frac{1}{2},0)+$	Reflection conditions

General:

16   *b*   1   (1) x,y,z (2) $\bar{x},\bar{y},z$ (3) $x+\frac{1}{4},\bar{y}+\frac{1}{4},z+\frac{1}{4}$ (4) $\bar{x}+\frac{1}{4},y+\frac{1}{4},z+\frac{1}{4}$

hkl : $h+k,h+l,k+l=2n$
$0kl$: $k+l=4n,\ k,l=2n$
$h0l$: $h+l=4n,\ h,l=2n$
$hk0$: $h,k=2n$
$h00$: $h=4n$
$0k0$: $k=4n$
$00l$: $l=4n$

Special: as above, plus

8   *a*   . . 2 $0,0,z$ $\frac{1}{4},\frac{1}{4},z+\frac{1}{4}$

hkl : $h=2n+1$
    or $h+k+l=4n$

Symmetry of special projections

Along [001] $p\,2\,g\,g$
$\mathbf{a}'=\frac{1}{2}\mathbf{a}$ $\mathbf{b}'=\frac{1}{2}\mathbf{b}$
Origin at $0,0,z$

Along [100] $c\,1\,m\,1$
$\mathbf{a}'=\frac{1}{2}\mathbf{b}$ $\mathbf{b}'=\frac{1}{2}\mathbf{c}$
Origin at $x,0,0$

Along [010] $c\,1\,1\,m$
$\mathbf{a}'=\frac{1}{2}\mathbf{c}$ $\mathbf{b}'=\frac{1}{2}\mathbf{a}$
Origin at $0,y,0$

Maximal non-isomorphic subgroups

I   [2] $F\,1\,d\,1\,(Cc,9)$ (1; 3)+
      [2] $F\,d\,1\,1\,(Cc,9)$ (1; 4)+
      [2] $F\,1\,1\,2\,(C2,5)$ (1; 2)+

IIa   none

IIb   none

Maximal isomorphic subgroups of lowest index

IIc   [3] $F\,d\,d\,2\,(\mathbf{a}'=3\mathbf{a}$ or $\mathbf{b}'=3\mathbf{b})\,(43)$; [3] $F\,d\,d\,2\,(\mathbf{c}'=3\mathbf{c})\,(43)$

Minimal non-isomorphic supergroups

I   [2] $F\,d\,d\,d\,(70)$; [2] $I\,4_1\,m\,d\,(109)$; [2] $I\,4_1\,c\,d\,(110)$; [2] $I\bar{4}\,2\,d\,(122)$

II   [2] $P\,n\,n\,2\,(\mathbf{a}'=\frac{1}{2}\mathbf{a},\mathbf{b}'=\frac{1}{2}\mathbf{b},\mathbf{c}'=\frac{1}{2}\mathbf{c})\,(34)$

$Pmna$ D_{2h}^7 mmm Orthorhombic

No. 53 $P\,2/m\,2/n\,2_1/a$

Patterson symmetry $Pmmm$

$P\frac{2}{m}\frac{2}{n}\frac{2_1}{a}$

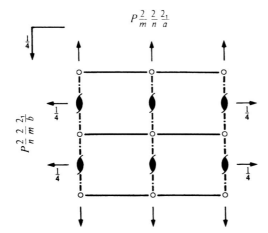

$P\frac{2}{m}\frac{2_1}{a}\frac{2}{n}$

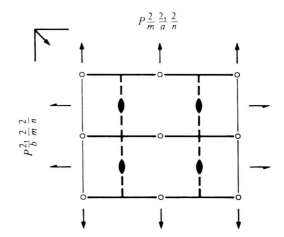

$P\frac{2_1}{c}\frac{2}{n}\frac{2}{m}$

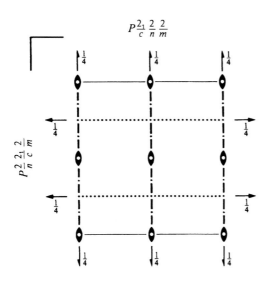

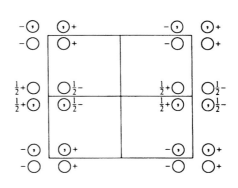

Origin at centre $(2/m)$ at $2/mn\,1$

Asymmetric unit $\quad 0 \le x \le \frac{1}{2}; \quad 0 \le y \le 1; \quad 0 \le z \le \frac{1}{4}$

Symmetry operations

(1) 1

(2) $2(0,0,\frac{1}{2}) \quad \frac{1}{4},0,z$

(3) $2 \quad \frac{1}{4},y,\frac{1}{4}$

(4) $2 \quad x,0,0$

(5) $\bar{1} \quad 0,0,0$

(6) $a \quad x,y,\frac{1}{4}$

(7) $n(\frac{1}{2},0,\frac{1}{2}) \quad x,0,z$

(8) $m \quad 0,y,z$

120

Generators selected   (1); $t(1,0,0)$; $t(0,1,0)$; $t(0,0,1)$; (2); (3); (5)

Positions

Multiplicity, Wyckoff letter, Site symmetry		Coordinates				Reflection conditions

General:

8   i 1   (1) x,y,z (2) $\bar{x}+\frac{1}{2},\bar{y},z+\frac{1}{2}$ (3) $\bar{x}+\frac{1}{2},y,\bar{z}+\frac{1}{2}$ (4) $x,\bar{y},\bar{z}$

　　　　　(5) $\bar{x},\bar{y},\bar{z}$ (6) $x+\frac{1}{2},y,\bar{z}+\frac{1}{2}$ (7) $x+\frac{1}{2},\bar{y},z+\frac{1}{2}$ (8) $\bar{x},y,z$

$h0l$: $h+l = 2n$
$hk0$: $h = 2n$
$h00$: $h = 2n$
$00l$: $l = 2n$

Special: as above, plus

4   h m . . $0,y,z$ $\frac{1}{2},\bar{y},z+\frac{1}{2}$ $\frac{1}{2},y,\bar{z}+\frac{1}{2}$ $0,\bar{y},\bar{z}$

no extra conditions

4   g . 2 . $\frac{1}{4},y,\frac{1}{4}$ $\frac{1}{4},\bar{y},\frac{3}{4}$ $\frac{3}{4},\bar{y},\frac{3}{4}$ $\frac{3}{4},y,\frac{1}{4}$

hkl : $h = 2n$

4   f 2 . . $x,\frac{1}{2},0$ $\bar{x}+\frac{1}{2},\frac{1}{2},\frac{1}{2}$ $\bar{x},\frac{1}{2},0$ $x+\frac{1}{2},\frac{1}{2},\frac{1}{2}$

hkl : $h+l = 2n$

4   e 2 . . $x,0,0$ $\bar{x}+\frac{1}{2},0,\frac{1}{2}$ $\bar{x},0,0$ $x+\frac{1}{2},0,\frac{1}{2}$

hkl : $h+l = 2n$

2   d $2/m$ . . $0,\frac{1}{2},0$ $\frac{1}{2},\frac{1}{2},\frac{1}{2}$

hkl : $h+l = 2n$

2   c $2/m$ . . $\frac{1}{2},\frac{1}{2},0$ $0,\frac{1}{2},\frac{1}{2}$

hkl : $h+l = 2n$

2   b $2/m$ . . $\frac{1}{2},0,0$ $0,0,\frac{1}{2}$

hkl : $h+l = 2n$

2   a $2/m$ . . $0,0,0$ $\frac{1}{2},0,\frac{1}{2}$

hkl : $h+l = 2n$

Symmetry of special projections

Along [001] $p2mm$ Along [100] $p2gm$ Along [010] $c2mm$
$\mathbf{a}' = \frac{1}{2}\mathbf{a}$ $\mathbf{b}' = \mathbf{b}$ $\mathbf{a}' = \mathbf{b}$ $\mathbf{b}' = \mathbf{c}$ $\mathbf{a}' = \mathbf{c}$ $\mathbf{b}' = \mathbf{a}$
Origin at $0,0,z$ Origin at $x,0,0$ Origin at $0,y,0$

Maximal non-isomorphic subgroups

I   [2] $Pmn2_1$ (31)   1; 2; 7; 8
　　[2] $P2na$ ($Pnc2$, 30)   1; 4; 6; 7
　　[2] $Pm2a$ ($Pma2$, 28)   1; 3; 6; 8
　　[2] $P222_1$ (17)   1; 2; 3; 4
　　[2] $P112_1/a$ ($P2_1/c$, 14)   1; 2; 5; 6
　　[2] $P12/n1$ ($P2/c$, 13)   1; 3; 5; 7
　　[2] $P2/m11$ ($P2/m$, 10)   1; 4; 5; 8

IIa   none

IIb   [2] $Pbna$ ($\mathbf{b}' = 2\mathbf{b}$) ($Pbcn$, 60); [2] $Pmnn$ ($\mathbf{b}' = 2\mathbf{b}$) ($Pnnm$, 58); [2] $Pbnn$ ($\mathbf{b}' = 2\mathbf{b}$) ($Pnna$, 52)

Maximal isomorphic subgroups of lowest index

IIc   [2] $Pmna$ ($\mathbf{b}' = 2\mathbf{b}$) (53); [3] $Pmna$ ($\mathbf{a}' = 3\mathbf{a}$) (53); [3] $Pmna$ ($\mathbf{c}' = 3\mathbf{c}$) (53)

Minimal non-isomorphic supergroups

I   none

II   [2] $Cmce$ (64); [2] $Bmmm$ ($Cmmm$, 65); [2] $Amaa$ ($Cccm$, 66); [2] $Imma$ (74); [2] $Pmaa$ ($\mathbf{c}' = \frac{1}{2}\mathbf{c}$) ($Pccm$, 49);
　　[2] $Pmcm$ ($\mathbf{a}' = \frac{1}{2}\mathbf{a}$) ($Pmma$, 51)

$P\frac{2_1}{n}\frac{2_1}{m}\frac{2_1}{a}$

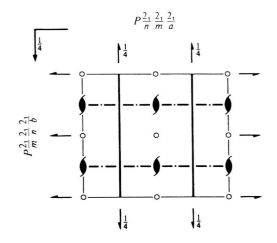

$P\frac{2_1}{n}\frac{2_1}{a}\frac{2_1}{m}$

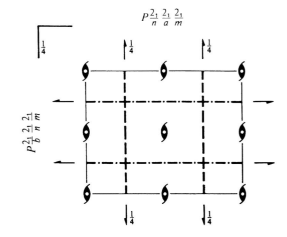

$P\frac{2_1}{c}\frac{2_1}{m}\frac{2_1}{n}$

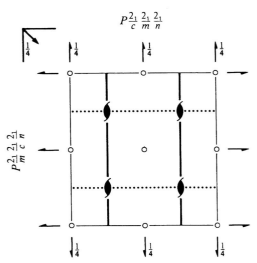

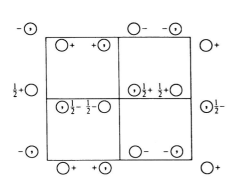

Origin at $\bar{1}$ on $12_1 1$

Asymmetric unit $\quad 0 \le x \le \frac{1}{2}; \quad 0 \le y \le \frac{1}{4}; \quad 0 \le z \le 1$

Symmetry operations

(1) 1

(2) $2(0,0,\frac{1}{2})\quad \frac{1}{4},0,z$

(3) $2(0,\frac{1}{2},0)\quad 0,y,0$

(4) $2(\frac{1}{2},0,0)\quad x,\frac{1}{4},\frac{1}{4}$

(5) $\bar{1}\quad 0,0,0$

(6) $a\quad x,y,\frac{1}{4}$

(7) $m\quad x,\frac{1}{4},z$

(8) $n(0,\frac{1}{2},\frac{1}{2})\quad \frac{1}{4},y,z$

Generators selected   (1); $t(1,0,0)$; $t(0,1,0)$; $t(0,0,1)$; (2); (3); (5)

Positions

Multiplicity, Wyckoff letter, Site symmetry	Coordinates				Reflection conditions

General:

8 d 1   (1) x,y,z (2) $\bar{x}+\frac{1}{2},\bar{y},z+\frac{1}{2}$ (3) $\bar{x},y+\frac{1}{2},\bar{z}$ (4) $x+\frac{1}{2},\bar{y}+\frac{1}{2},\bar{z}+\frac{1}{2}$ $0kl$: $k+l=2n$
   (5) $\bar{x},\bar{y},\bar{z}$ (6) $x+\frac{1}{2},y,\bar{z}+\frac{1}{2}$ (7) $x,\bar{y}+\frac{1}{2},z$ (8) $\bar{x}+\frac{1}{2},y+\frac{1}{2},z+\frac{1}{2}$ $hk0$: $h=2n$
$h00$: $h=2n$
$0k0$: $k=2n$
$00l$: $l=2n$

Special: as above, plus

4 c $.m.$ $x,\frac{1}{4},z$ $\bar{x}+\frac{1}{2},\frac{3}{4},z+\frac{1}{2}$ $\bar{x},\frac{3}{4},\bar{z}$ $x+\frac{1}{2},\frac{1}{4},\bar{z}+\frac{1}{2}$ no extra conditions

4 b $\bar{1}$ $0,0,\frac{1}{2}$ $\frac{1}{2},0,0$ $0,\frac{1}{2},\frac{1}{2}$ $\frac{1}{2},\frac{1}{2},0$ hkl : $h+l,k=2n$

4 a $\bar{1}$ $0,0,0$ $\frac{1}{2},0,\frac{1}{2}$ $0,\frac{1}{2},0$ $\frac{1}{2},\frac{1}{2},\frac{1}{2}$ hkl : $h+l,k=2n$

Symmetry of special projections

Along [001] $p2gm$ Along [100] $c2mm$ Along [010] $p2gg$
$\mathbf{a}'=\frac{1}{2}\mathbf{a}$ $\mathbf{b}'=\mathbf{b}$ $\mathbf{a}'=\mathbf{b}$ $\mathbf{b}'=\mathbf{c}$ $\mathbf{a}'=\mathbf{c}$ $\mathbf{b}'=\mathbf{a}$
Origin at $0,0,z$ Origin at $x,\frac{1}{4},\frac{1}{4}$ Origin at $0,y,0$

Maximal non-isomorphic subgroups

I   [2] $Pn2_1a\,(Pna2_1,\,33)$ 1; 3; 6; 8
[2] $Pnm2_1\,(Pmn2_1,\,31)$ 1; 2; 7; 8
[2] $P2_1ma\,(Pmc2_1,\,26)$ 1; 4; 6; 7
[2] $P2_12_12_1\,(19)$ 1; 2; 3; 4
[2] $P112_1/a\,(P2_1/c,\,14)$ 1; 2; 5; 6
[2] $P2_1/n11\,(P2_1/c,\,14)$ 1; 4; 5; 8
[2] $P12_1/m1\,(P2_1/m,\,11)$ 1; 3; 5; 7

IIa   none
IIb   none

Maximal isomorphic subgroups of lowest index
IIc   [3] $Pnma\,(\mathbf{a}'=3\mathbf{a})\,(62)$; [3] $Pnma\,(\mathbf{b}'=3\mathbf{b})\,(62)$; [3] $Pnma\,(\mathbf{c}'=3\mathbf{c})\,(62)$

Minimal non-isomorphic supergroups
I   none
II   [2] $Amma\,(Cmcm,\,63)$; [2] $Bbmm\,(Cmcm,\,63)$; [2] $Ccme\,(Cmce,\,64)$; [2] $Imma\,(74)$; [2] $Pcma\,(\mathbf{b}'=\frac{1}{2}\mathbf{b})\,(Pbam,\,55)$;
[2] $Pbma\,(\mathbf{c}'=\frac{1}{2}\mathbf{c})\,(Pbcm,\,57)$; [2] $Pnmm\,(\mathbf{a}'=\frac{1}{2}\mathbf{a})\,(Pmmn,\,59)$

Cmce

D_{2h}^{18}

mmm

Orthorhombic

No. 64

$C\ 2/m\ 2/c\ 2_1/e$

Patterson symmetry *Cmmm*

Former space-group symbol *Cmca*; *cf.* Chapter 1.3

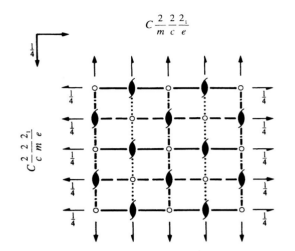

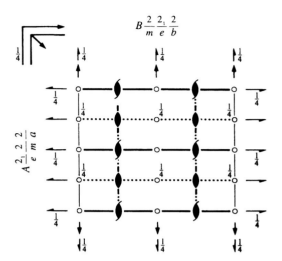

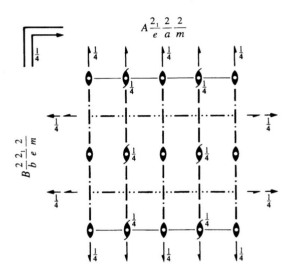

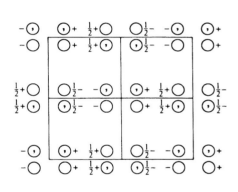

Origin at centre $(2/m)$ at $2/mn\,1$

Asymmetric unit $0 \le x \le \frac{1}{4};\quad 0 \le y \le \frac{1}{2};\quad 0 \le z \le \frac{1}{2}$

Symmetry operations

For $(0,0,0)+$ set

(1) 1

(2) $2(0,0,\frac{1}{2})\quad 0,\frac{1}{4},z$

(3) $2(0,\frac{1}{2},0)\quad 0,y,\frac{1}{4}$

(4) $2\quad x,0,0$

(5) $\bar{1}\quad 0,0,0$

(6) $b\quad x,y,\frac{1}{4}$

(7) $c\quad x,\frac{1}{4},z$

(8) $m\quad 0,y,z$

For $(\frac{1}{2},\frac{1}{2},0)+$ set

(1) $t(\frac{1}{2},\frac{1}{2},0)$

(2) $2(0,0,\frac{1}{2})\quad \frac{1}{4},0,z$

(3) $2\quad \frac{1}{4},y,\frac{1}{4}$

(4) $2(\frac{1}{2},0,0)\quad x,\frac{1}{4},0$

(5) $\bar{1}\quad \frac{1}{4},\frac{1}{4},0$

(6) $a\quad x,y,\frac{1}{4}$

(7) $n(\frac{1}{2},0,\frac{1}{2})\quad x,0,z$

(8) $b\quad \frac{1}{4},y,z$

Generators selected   (1); $t(1,0,0)$; $t(0,1,0)$; $t(0,0,1)$; $t(\frac{1}{2},\frac{1}{2},0)$; (2); (3); (5)

Positions

Multiplicity,
Wyckoff letter,
Site symmetry

Coordinates

$(0,0,0)+ \quad (\frac{1}{2},\frac{1}{2},0)+$

Reflection conditions

General:

16 g 1	(1) x,y,z	(2) $\bar{x},\bar{y}+\frac{1}{2},z+\frac{1}{2}$	(3) $\bar{x},y+\frac{1}{2},\bar{z}+\frac{1}{2}$	(4) $x,\bar{y},\bar{z}$
	(5) $\bar{x},\bar{y},\bar{z}$	(6) $x,y+\frac{1}{2},\bar{z}+\frac{1}{2}$	(7) $x,\bar{y}+\frac{1}{2},z+\frac{1}{2}$	(8) $\bar{x},y,z$

$hkl:\ h+k=2n$
$0kl:\ k=2n$
$h0l:\ h,l=2n$
$hk0:\ h,k=2n$
$h00:\ h=2n$
$0k0:\ k=2n$
$00l:\ l=2n$

Special: as above, plus

8	f	m . .	$0,y,z$	$0,\bar{y}+\frac{1}{2},z+\frac{1}{2}$	$0,y+\frac{1}{2},\bar{z}+\frac{1}{2}$ $0,\bar{y},\bar{z}$

no extra conditions

8	e	. 2 .	$\frac{1}{4},y,\frac{1}{4}$	$\frac{3}{4},\bar{y}+\frac{1}{2},\frac{3}{4}$	$\frac{3}{4},\bar{y},\frac{3}{4}$ $\frac{1}{4},y+\frac{1}{2},\frac{1}{4}$

$hkl:\ h=2n$

8	d	2 . .	$x,0,0$	$\bar{x},\frac{1}{2},\frac{1}{2}$	$\bar{x},0,0$ $x,\frac{1}{2},\frac{1}{2}$

$hkl:\ k+l=2n$

8	c	$\bar{1}$	$\frac{1}{4},\frac{1}{4},0$	$\frac{3}{4},\frac{1}{4},\frac{1}{2}$	$\frac{3}{4},\frac{3}{4},\frac{1}{2}$ $\frac{1}{4},\frac{3}{4},0$

$hkl:\ k,l=2n$

4	b	$2/m$ . .	$\frac{1}{2},0,0$	$\frac{1}{2},\frac{1}{2},\frac{1}{2}$

$hkl:\ k+l=2n$

4	a	$2/m$ . .	$0,0,0$	$0,\frac{1}{2},\frac{1}{2}$

$hkl:\ k+l=2n$

Symmetry of special projections

Along [001] $p2mm$
$\mathbf{a}'=\frac{1}{2}\mathbf{a}$ $\mathbf{b}'=\frac{1}{2}\mathbf{b}$
Origin at $0,0,z$

Along [100] $p2gm$
$\mathbf{a}'=\frac{1}{2}\mathbf{b}$ $\mathbf{b}'=\mathbf{c}$
Origin at $x,0,0$

Along [010] $p2mm$
$\mathbf{a}'=\frac{1}{2}\mathbf{c}$ $\mathbf{b}'=\frac{1}{2}\mathbf{a}$
Origin at $0,y,0$

Maximal non-isomorphic subgroups

I	[2] $C2ce\,(Aea2,\,41)$	$(1;\ 4;\ 6;\ 7)+$
	[2] $Cm2e\,(Aem2,\,39)$	$(1;\ 3;\ 6;\ 8)+$
	[2] $Cmc2_1\,(36)$	$(1;\ 2;\ 7;\ 8)+$
	[2] $C222_1\,(20)$	$(1;\ 2;\ 3;\ 4)+$
	[2] $C12/c1\,(C2/c,\,15)$	$(1;\ 3;\ 5;\ 7)+$
	[2] $C112_1/e\,(P2_1/c,\,14)$	$(1;\ 2;\ 5;\ 6)+$
	[2] $C2/m11\,(C2/m,\,12)$	$(1;\ 4;\ 5;\ 8)+$
IIa	[2] $Pmnb\,(Pnma,\,62)$	$1;\ 3;\ 6;\ 8;\ (2;\ 4;\ 5;\ 7)+(\frac{1}{2},\frac{1}{2},0)$
	[2] $Pbca\,(61)$	$1;\ 3;\ 5;\ 7;\ (2;\ 4;\ 6;\ 8)+(\frac{1}{2},\frac{1}{2},0)$
	[2] $Pbna\,(Pbcn,\,60)$	$1;\ 2;\ 3;\ 4;\ (5;\ 6;\ 7;\ 8)+(\frac{1}{2},\frac{1}{2},0)$
	[2] $Pmca\,(Pbcm,\,57)$	$1;\ 2;\ 7;\ 8;\ (3;\ 4;\ 5;\ 6)+(\frac{1}{2},\frac{1}{2},0)$
	[2] $Pbnb\,(Pccn,\,56)$	$1;\ 2;\ 5;\ 6;\ (3;\ 4;\ 7;\ 8)+(\frac{1}{2},\frac{1}{2},0)$
	[2] $Pmcb\,(Pbam,\,55)$	$1;\ 2;\ 3;\ 4;\ 5;\ 6;\ 7;\ 8$
	[2] $Pbcb\,(Pcca,\,54)$	$1;\ 4;\ 6;\ 7;\ (2;\ 3;\ 5;\ 8)+(\frac{1}{2},\frac{1}{2},0)$
	[2] $Pmna\,(53)$	$1;\ 4;\ 5;\ 8;\ (2;\ 3;\ 6;\ 7)+(\frac{1}{2},\frac{1}{2},0)$
IIb	none	

Maximal isomorphic subgroups of lowest index

IIc   [3] $Cmce\,(\mathbf{a}'=3\mathbf{a})\,(64)$; [3] $Cmce\,(\mathbf{b}'=3\mathbf{b})\,(64)$; [3] $Cmce\,(\mathbf{c}'=3\mathbf{c})\,(64)$

Minimal non-isomorphic supergroups

I   none

II   [2] $Fmmm\,(69)$; [2] $Pmcm\,(\mathbf{a}'=\frac{1}{2}\mathbf{a},\mathbf{b}'=\frac{1}{2}\mathbf{b})\,(Pmma,\,51)$; [2] $Cmme\,(\mathbf{c}'=\frac{1}{2}\mathbf{c})\,(67)$

Ibca

D_{2h}^{27}

mmm

Orthorhombic

No. 73

$I\ 2_1/b\ 2_1/c\ 2_1/a$

Patterson symmetry *Immm*

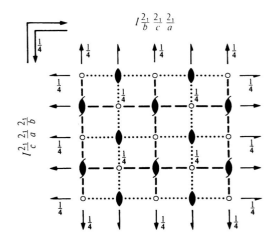

$I\frac{2_1}{b}\frac{2_1}{c}\frac{2_1}{a}$

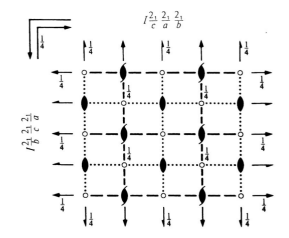

$I\frac{2_1}{c}\frac{2_1}{a}\frac{2_1}{b}$

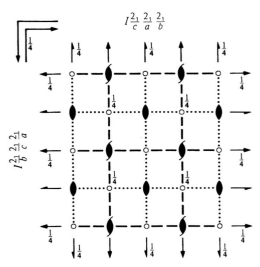

$I\frac{2_1}{c}\frac{2_1}{a}\frac{2_1}{b}$

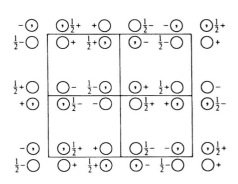

Origin at $\bar{1}$ at cab

Asymmetric unit $\quad 0 \le x \le \frac{1}{4}; \quad 0 \le y \le \frac{1}{2}; \quad 0 \le z \le \frac{1}{2}$

Symmetry operations

For $(0,0,0)+$ set

(1) 1
(2) $2(0,0,\frac{1}{2}) \quad \frac{1}{4},0,z$
(3) $2(0,\frac{1}{2},0) \quad 0,y,\frac{1}{4}$
(4) $2(\frac{1}{2},0,0) \quad x,\frac{1}{4},0$
(5) $\bar{1} \quad 0,0,0$
(6) $a \quad x,y,\frac{1}{4}$
(7) $c \quad x,\frac{1}{4},z$
(8) $b \quad \frac{1}{4},y,z$

For $(\frac{1}{2},\frac{1}{2},\frac{1}{2})+$ set

(1) $t(\frac{1}{2},\frac{1}{2},\frac{1}{2})$
(2) $2 \quad 0,\frac{1}{4},z$
(3) $2 \quad \frac{1}{4},y,0$
(4) $2 \quad x,0,\frac{1}{4}$
(5) $\bar{1} \quad \frac{1}{4},\frac{1}{4},\frac{1}{4}$
(6) $b \quad x,y,0$
(7) $a \quad x,0,z$
(8) $c \quad 0,y,z$

Generators selected   (1); $t(1,0,0)$; $t(0,1,0)$; $t(0,0,1)$; $t(\frac{1}{2},\frac{1}{2},\frac{1}{2})$; (2); (3); (5)

Positions

Multiplicity, Wyckoff letter, Site symmetry	Coordinates $(0,0,0)+$ $(\frac{1}{2},\frac{1}{2},\frac{1}{2})+$				Reflection conditions

General:

16	f	1	(1) x,y,z	(2) $\bar{x}+\frac{1}{2},\bar{y},z+\frac{1}{2}$	(3) $\bar{x},y+\frac{1}{2},\bar{z}+\frac{1}{2}$	(4) $x+\frac{1}{2},\bar{y}+\frac{1}{2},\bar{z}$
			(5) $\bar{x},\bar{y},\bar{z}$	(6) $x+\frac{1}{2},y,\bar{z}+\frac{1}{2}$	(7) $x,\bar{y}+\frac{1}{2},z+\frac{1}{2}$	(8) $\bar{x}+\frac{1}{2},y+\frac{1}{2},z$

hkl : $h+k+l=2n$
$0kl$: $k,l=2n$
$h0l$: $h,l=2n$
$hk0$: $h,k=2n$
$h00$: $h=2n$
$0k0$: $k=2n$
$00l$: $l=2n$

Special: as above, plus

8	e	..2	$0,\frac{1}{4},z$	$0,\frac{3}{4},\bar{z}+\frac{1}{2}$	$0,\frac{3}{4},\bar{z}$	$0,\frac{1}{4},z+\frac{1}{2}$	hkl : $l=2n$
8	d	.2.	$\frac{1}{4},y,0$	$\frac{1}{4},\bar{y},\frac{1}{2}$	$\frac{3}{4},\bar{y},0$	$\frac{3}{4},y,\frac{1}{2}$	hkl : $k=2n$
8	c	2..	$x,0,\frac{1}{4}$	$\bar{x}+\frac{1}{2},0,\frac{3}{4}$	$\bar{x},0,\frac{3}{4}$	$x+\frac{1}{2},0,\frac{1}{4}$	hkl : $h=2n$
8	b	$\bar{1}$	$\frac{1}{4},\frac{1}{4},\frac{1}{4}$	$\frac{1}{4},\frac{3}{4},\frac{3}{4}$	$\frac{3}{4},\frac{3}{4},\frac{1}{4}$	$\frac{3}{4},\frac{1}{4},\frac{3}{4}$	hkl : $k,l=2n$
8	a	$\bar{1}$	$0,0,0$	$\frac{1}{2},0,\frac{1}{2}$	$0,\frac{1}{2},\frac{1}{2}$	$\frac{1}{2},\frac{1}{2},0$	hkl : $k,l=2n$

Symmetry of special projections

Along $[001]$ $p2mm$	Along $[100]$ $p2mm$	Along $[010]$ $p2mm$
$\mathbf{a}'=\frac{1}{2}\mathbf{a}$ $\mathbf{b}'=\frac{1}{2}\mathbf{b}$	$\mathbf{a}'=\frac{1}{2}\mathbf{b}$ $\mathbf{b}'=\frac{1}{2}\mathbf{c}$	$\mathbf{a}'=\frac{1}{2}\mathbf{c}$ $\mathbf{b}'=\frac{1}{2}\mathbf{a}$
Origin at $0,0,z$	Origin at $x,0,0$	Origin at $0,y,0$

Maximal non-isomorphic subgroups

I	$[2]Ibc2(Iba2,45)$	$(1;\ 2;\ 7;\ 8)+$
	$[2]Ib2a(Iba2,45)$	$(1;\ 3;\ 6;\ 8)+$
	$[2]I2ca(Iba2,45)$	$(1;\ 4;\ 6;\ 7)+$
	$[2]I2_12_12_1(24)$	$(1;\ 2;\ 3;\ 4)+$
	$[2]I112/a(C2/c,15)$	$(1;\ 2;\ 5;\ 6)+$
	$[2]I12/c1(C2/c,15)$	$(1;\ 3;\ 5;\ 7)+$
	$[2]I2/b11(C2/c,15)$	$(1;\ 4;\ 5;\ 8)+$

IIa	$[2]Pbca(61)$	$1;\ 2;\ 3;\ 4;\ 5;\ 6;\ 7;\ 8$
	$[2]Pcab(Pbca,61)$	$1;\ 2;\ 3;\ 4;\ (5;\ 6;\ 7;\ 8)+(\frac{1}{2},\frac{1}{2},\frac{1}{2})$
	$[2]Pcaa(Pcca,54)$	$1;\ 2;\ 5;\ 6;\ (3;\ 4;\ 7;\ 8)+(\frac{1}{2},\frac{1}{2},\frac{1}{2})$
	$[2]Pccb(Pcca,54)$	$1;\ 3;\ 5;\ 7;\ (2;\ 4;\ 6;\ 8)+(\frac{1}{2},\frac{1}{2},\frac{1}{2})$
	$[2]Pbab(Pcca,54)$	$1;\ 4;\ 5;\ 8;\ (2;\ 3;\ 6;\ 7)+(\frac{1}{2},\frac{1}{2},\frac{1}{2})$
	$[2]Pbcb(Pcca,54)$	$1;\ 2;\ 7;\ 8;\ (3;\ 4;\ 5;\ 6)+(\frac{1}{2},\frac{1}{2},\frac{1}{2})$
	$[2]Pbaa(Pcca,54)$	$1;\ 3;\ 6;\ 8;\ (2;\ 4;\ 5;\ 7)+(\frac{1}{2},\frac{1}{2},\frac{1}{2})$
	$[2]Pcca(54)$	$1;\ 4;\ 6;\ 7;\ (2;\ 3;\ 5;\ 8)+(\frac{1}{2},\frac{1}{2},\frac{1}{2})$

IIb   none

Maximal isomorphic subgroups of lowest index

IIc $[3]Ibca\,(\mathbf{a}'=3\mathbf{a}$ or $\mathbf{b}'=3\mathbf{b}$ or $\mathbf{c}'=3\mathbf{c})\,(73)$

Minimal non-isomorphic supergroups

I $[2]I4_1/acd\,(142);\ [3]Ia\bar{3}\,(206)$

II $[2]Aemm\,(\mathbf{a}'=\frac{1}{2}\mathbf{a})\,(Cmme,67);\ [2]Bmem\,(\mathbf{b}'=\frac{1}{2}\mathbf{b})\,(Cmme,67);\ [2]Cmme\,(\mathbf{c}'=\frac{1}{2}\mathbf{c})\,(67)$

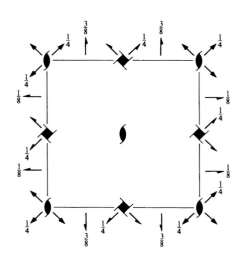
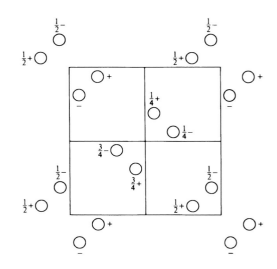

Origin on 2 [1 1 0] at 2_1 1 (1, 2)

Asymmetric unit $0 \le x \le 1$; $0 \le y \le 1$; $0 \le z \le \frac{1}{8}$

Symmetry operations

(1) 1

(2) $2(0,0,\frac{1}{2})$ $0,0,z$

(3) $4^+(0,0,\frac{1}{4})$ $0,\frac{1}{2},z$

(4) $4^-(0,0,\frac{3}{4})$ $\frac{1}{2},0,z$

(5) $2(0,\frac{1}{2},0)$ $\frac{1}{4},y,\frac{1}{8}$

(6) $2(\frac{1}{2},0,0)$ $x,\frac{1}{4},\frac{3}{8}$

(7) 2 $x,x,0$

(8) 2 $x,\bar{x},\frac{1}{4}$

Generators selected   (1); $t(1,0,0)$; $t(0,1,0)$; $t(0,0,1)$; $t(\frac{1}{2},\frac{1}{2},\frac{1}{2})$; (2); (3); (5); (9)

Positions

Multiplicity, Wyckoff letter, Site symmetry	Coordinates $(0,0,0)+$ $(\frac{1}{2},\frac{1}{2},\frac{1}{2})+$	Reflection conditions

General:

32 i 1	(1) x,y,z (2) $\bar{x}+\frac{1}{2},\bar{y},z+\frac{1}{2}$ (3) $\bar{y}+\frac{1}{4},x+\frac{3}{4},z+\frac{1}{4}$ (4) $y+\frac{1}{4},\bar{x}+\frac{1}{4},z+\frac{3}{4}$	$hkl:\ h+k+l=2n$

(5) $\bar{x}+\frac{1}{4},y,\bar{z}+\frac{1}{2}$ (6) $x,\bar{y},\bar{z}$ (7) $y+\frac{1}{4},x+\frac{3}{4},\bar{z}+\frac{1}{4}$ (8) $\bar{y}+\frac{1}{4},\bar{x}+\frac{1}{4},\bar{z}+\frac{3}{4}$

(9) $\bar{x},\bar{y},\bar{z}$ (10) $x+\frac{1}{2},y,\bar{z}+\frac{1}{2}$ (11) $y+\frac{3}{4},\bar{x}+\frac{1}{4},\bar{z}+\frac{3}{4}$ (12) $\bar{y}+\frac{3}{4},x+\frac{3}{4},\bar{z}+\frac{1}{4}$

(13) $x+\frac{1}{2},\bar{y},z+\frac{1}{2}$ (14) $\bar{x},y,z$ (15) $\bar{y}+\frac{3}{4},\bar{x}+\frac{1}{4},z+\frac{3}{4}$ (16) $y+\frac{3}{4},x+\frac{3}{4},z+\frac{1}{4}$

Reflection conditions (General):
$hkl:\ h+k+l=2n$
$hk0:\ h,k=2n$
$0kl:\ k+l=2n$
$hhl:\ 2h+l=4n$
$00l:\ l=4n$
$h00:\ h=2n$
$h\bar{h}0:\ h=2n$

Special: as above, plus

| 16 h $.m.$ | $0,y,z$ $\frac{1}{2},\bar{y},z+\frac{1}{2}$ $\bar{y}+\frac{1}{4},\frac{3}{4},z+\frac{1}{4}$ $y+\frac{1}{4},\frac{1}{4},z+\frac{3}{4}$ | no extra conditions |
| | $\frac{1}{2},y,\bar{z}+\frac{1}{2}$ $0,\bar{y},\bar{z}$ $y+\frac{1}{4},\frac{3}{4},\bar{z}+\frac{1}{4}$ $\bar{y}+\frac{1}{4},\frac{1}{4},\bar{z}+\frac{3}{4}$ | |

| 16 g $..2$ | $x,x+\frac{1}{4},\frac{7}{8}$ $\bar{x}+\frac{1}{2},\bar{x}+\frac{3}{4},\frac{3}{8}$ $\bar{x},x+\frac{3}{4},\frac{1}{8}$ $x+\frac{1}{2},\bar{x}+\frac{1}{4},\frac{5}{8}$ | $hkl:\ l=2n+1$ |
| | $\bar{x},\bar{x}+\frac{3}{4},\frac{1}{8}$ $x+\frac{1}{2},x+\frac{1}{4},\frac{5}{8}$ $x,\bar{x}+\frac{1}{4},\frac{7}{8}$ $\bar{x}+\frac{1}{2},x+\frac{3}{4},\frac{3}{8}$ | or $2h+l=4n$ |

| 16 f $.2.$ | $x,0,0$ $\bar{x}+\frac{1}{2},0,\frac{1}{2}$ $\frac{1}{4},x+\frac{3}{4},\frac{1}{4}$ $\frac{1}{4},\bar{x}+\frac{1}{4},\frac{3}{4}$ | $hkl:\ l=2n+1$ |
| | $\bar{x},0,0$ $x+\frac{1}{2},0,\frac{1}{2}$ $\frac{3}{4},\bar{x}+\frac{1}{4},\frac{3}{4}$ $\frac{3}{4},x+\frac{3}{4},\frac{1}{4}$ | or $h=2n$ |

| 8 e $2mm.$ | $0,\frac{1}{4},z$ $0,\frac{3}{4},z+\frac{1}{4}$ $\frac{1}{2},\frac{1}{4},\bar{z}+\frac{1}{2}$ $\frac{1}{2},\frac{3}{4},\bar{z}+\frac{1}{4}$ | $hkl:\ l=2n+1$ or $2h+l=4n$ |

| 8 d $.2/m.$ | $0,0,\frac{1}{2}$ $\frac{1}{2},0,0$ $\frac{1}{4},\frac{3}{4},\frac{3}{4}$ $\frac{1}{4},\frac{1}{4},\frac{1}{4}$ | $hkl:\ l=2n+1$ |
| 8 c $.2/m.$ | $0,0,0$ $\frac{1}{2},0,\frac{1}{2}$ $\frac{1}{4},\frac{3}{4},\frac{1}{4}$ $\frac{1}{4},\frac{1}{4},\frac{3}{4}$ | or $h,k=2n,\ h+k+l=4n$ |

| 4 b $\bar{4}m2$ | $0,\frac{1}{4},\frac{3}{8}$ $0,\frac{3}{4},\frac{5}{8}$ | $hkl:\ l=2n+1$ |
| 4 a $\bar{4}m2$ | $0,\frac{3}{4},\frac{1}{8}$ $\frac{1}{2},\frac{3}{4},\frac{3}{8}$ | or $2h+l=4n$ |

Symmetry of special projections

Along [001] $p4mm$
$\mathbf{a}'=\frac{1}{2}\mathbf{a}$ $\mathbf{b}'=\frac{1}{2}\mathbf{b}$
Origin at $\frac{1}{4},0,z$

Along [100] $c2mm$
$\mathbf{a}'=\mathbf{b}$ $\mathbf{b}'=\mathbf{c}$
Origin at $x,\frac{1}{4},\frac{1}{4}$

Along [110] $c2mm$
$\mathbf{a}'=\frac{1}{2}(-\mathbf{a}+\mathbf{b})$ $\mathbf{b}'=\frac{1}{2}\mathbf{c}$
Origin at $x,x+\frac{1}{4},\frac{1}{8}$

Maximal non-isomorphic subgroups

I   [2] $I\bar{4}2d$ (122)   (1; 2; 5; 6; 11; 12; 15; 16)+
     [2] $I\bar{4}m2$ (119)   (1; 2; 7; 8; 11; 12; 13; 14)+
     [2] $I4_1md$ (109)   (1; 2; 3; 4; 13; 14; 15; 16)+
     [2] $I4_122$ (98)   (1; 2; 3; 4; 5; 6; 7; 8)+
     [2] $I4_1/a11$ ($I4_1/a$, 88)   (1; 2; 3; 4; 9; 10; 11; 12)+
     [2] $I2/a2/m1$ ($Imma$, 74)   (1; 2; 5; 6; 9; 10; 13; 14)+
     [2] $I2/a12/d$ ($Fddd$, 70)   (1; 2; 7; 8; 9; 10; 15; 16)+

IIa   none
IIb   none

Maximal isomorphic subgroups of lowest index

IIc   [3] $I4_1/amd$ ($\mathbf{c}'=3\mathbf{c}$) (141); [9] $I4_1/amd$ ($\mathbf{a}'=3\mathbf{a},\mathbf{b}'=3\mathbf{b}$) (141)

Minimal non-isomorphic supergroups

I   [3] $Fd\bar{3}m$ (227)
II   [2] $C4_2/amd$ ($\mathbf{c}'=\frac{1}{2}\mathbf{c}$) ($P4_2/nnm$, 134)

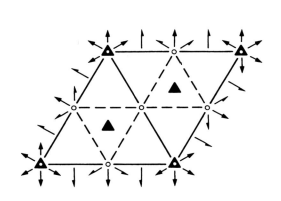

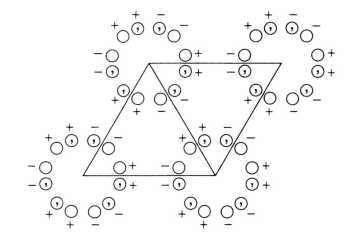

Origin at centre ($\bar{3}1m$)

Asymmetric unit $0 \le x \le \frac{2}{3}$; $0 \le y \le \frac{1}{2}$; $0 \le z \le \frac{1}{2}$; $x \le (1+y)/2$; $y \le \min(1-x,x)$

Vertices $0,0,0$ $\frac{1}{2},0,0$ $\frac{2}{3},\frac{1}{3},0$ $\frac{1}{2},\frac{1}{2},0$

$0,0,\frac{1}{2}$ $\frac{1}{2},0,\frac{1}{2}$ $\frac{2}{3},\frac{1}{3},\frac{1}{2}$ $\frac{1}{2},\frac{1}{2},\frac{1}{2}$

Symmetry operations

(1) 1
(2) 3^+ $0,0,z$
(3) 3^- $0,0,z$
(4) 2 $x,\bar{x},0$
(5) 2 $x,2x,0$
(6) 2 $2x,x,0$
(7) $\bar{1}$ $0,0,0$
(8) $\bar{3}^+$ $0,0,z$; $0,0,0$
(9) $\bar{3}^-$ $0,0,z$; $0,0,0$
(10) m x,x,z
(11) m $x,0,z$
(12) m $0,y,z$

Generators selected   (1); $t(1,0,0)$; $t(0,1,0)$; $t(0,0,1)$; (2); (4); (7)

Positions

Multiplicity, Wyckoff letter, Site symmetry			Coordinates		Reflection conditions

General:

12 l 1   (1) x,y,z (2) $\bar{y},x-y,z$ (3) $\bar{x}+y,\bar{x},z$ no conditions
              (4) $\bar{y},\bar{x},\bar{z}$ (5) $\bar{x}+y,y,\bar{z}$ (6) $x,x-y,\bar{z}$
              (7) $\bar{x},\bar{y},\bar{z}$ (8) $y,\bar{x}+y,\bar{z}$ (9) $x-y,x,\bar{z}$
              (10) y,x,z (11) $x-y,\bar{y},z$ (12) $\bar{x},\bar{x}+y,z$

Special: no extra conditions

6 k $..m$ $x,0,z$ $0,x,z$ $\bar{x},\bar{x},z$ $0,\bar{x},\bar{z}$ $\bar{x},0,\bar{z}$ $x,x,\bar{z}$

6 j $..2$ $x,\bar{x},\frac{1}{2}$ $x,2x,\frac{1}{2}$ $2\bar{x},\bar{x},\frac{1}{2}$ $\bar{x},x,\frac{1}{2}$ $\bar{x},2\bar{x},\frac{1}{2}$ $2x,x,\frac{1}{2}$

6 i $..2$ $x,\bar{x},0$ $x,2x,0$ $2\bar{x},\bar{x},0$ $\bar{x},x,0$ $\bar{x},2\bar{x},0$ $2x,x,0$

4 h $3..$ $\frac{1}{3},\frac{2}{3},z$ $\frac{1}{3},\frac{2}{3},\bar{z}$ $\frac{2}{3},\frac{1}{3},\bar{z}$ $\frac{2}{3},\frac{1}{3},z$

3 g $..2/m$ $\frac{1}{2},0,\frac{1}{2}$ $0,\frac{1}{2},\frac{1}{2}$ $\frac{1}{2},\frac{1}{2},\frac{1}{2}$

3 f $..2/m$ $\frac{1}{2},0,0$ $0,\frac{1}{2},0$ $\frac{1}{2},\frac{1}{2},0$

2 e $3.m$ $0,0,z$ $0,0,\bar{z}$

2 d 3.2 $\frac{1}{3},\frac{2}{3},\frac{1}{2}$ $\frac{2}{3},\frac{1}{3},\frac{1}{2}$

2 c 3.2 $\frac{1}{3},\frac{2}{3},0$ $\frac{2}{3},\frac{1}{3},0$

1 b $\bar{3}.m$ $0,0,\frac{1}{2}$

1 a $\bar{3}.m$ $0,0,0$

Symmetry of special projections

Along [001] $p6mm$
$\mathbf{a}'=\mathbf{a}$ $\mathbf{b}'=\mathbf{b}$
Origin at $0,0,z$

Along [100] $p2mm$
$\mathbf{a}'=\frac{1}{2}(\mathbf{a}+2\mathbf{b})$ $\mathbf{b}'=\mathbf{c}$
Origin at $x,0,0$

Along [210] $p2$
$\mathbf{a}'=\frac{1}{2}\mathbf{b}$ $\mathbf{b}'=\mathbf{c}$
Origin at $x,\frac{1}{2}x,0$

Maximal non-isomorphic subgroups

I   [2] $P31m$ (157)   1; 2; 3; 10; 11; 12
      [2] $P312$ (149)   1; 2; 3; 4; 5; 6
      [2] $P\bar{3}11$ ($P\bar{3}$, 147)   1; 2; 3; 7; 8; 9
      $\left\{\begin{array}{l}\end{array}\right.$ [3] $P112/m$ ($C2/m$, 12)   1; 4; 7; 10
      [3] $P112/m$ ($C2/m$, 12)   1; 5; 7; 11
      [3] $P112/m$ ($C2/m$, 12)   1; 6; 7; 12

IIa   none

IIb   [2] $P\bar{3}1c$ ($\mathbf{c}'=2\mathbf{c}$) (163); [3] $H\bar{3}1m$ ($\mathbf{a}'=3\mathbf{a},\mathbf{b}'=3\mathbf{b}$) ($P\bar{3}m1$, 164); [3] $R\bar{3}m$ ($\mathbf{a}'=\mathbf{a}-\mathbf{b},\mathbf{b}'=\mathbf{a}+2\mathbf{b},\mathbf{c}'=3\mathbf{c}$) (166); [3] $R\bar{3}m$ ($\mathbf{a}'=2\mathbf{a}+\mathbf{b},\mathbf{b}'=-\mathbf{a}+\mathbf{b},\mathbf{c}'=3\mathbf{c}$) (166)

Maximal isomorphic subgroups of lowest index

IIc   [2] $P\bar{3}1m$ ($\mathbf{c}'=2\mathbf{c}$) (162); [4] $P\bar{3}1m$ ($\mathbf{a}'=2\mathbf{a},\mathbf{b}'=2\mathbf{b}$) (162)

Minimal non-isomorphic supergroups

I   [2] $P6/mmm$ (191); [2] $P6_3/mcm$ (193)

II   [3] $H\bar{3}1m$ ($P\bar{3}m1$, 164)

$P\bar{3}m1$ D_{3d}^{3} $\bar{3}m1$ Trigonal

No. 164 $P\bar{3}2/m1$

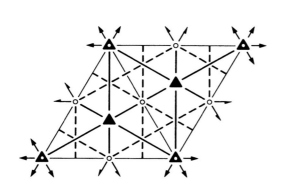
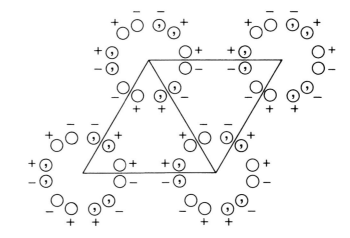

Origin at centre $(\bar{3}m1)$

Asymmetric unit $0 \le x \le \frac{2}{3};\quad 0 \le y \le \frac{1}{3};\quad 0 \le z \le 1;\quad x \le (1+y)/2;\quad y \le x/2$

Vertices $0,0,0\quad \frac{1}{2},0,0\quad \frac{2}{3},\frac{1}{3},0$

$0,0,1\quad \frac{1}{2},0,1\quad \frac{2}{3},\frac{1}{3},1$

Symmetry operations

(1) 1
(2) 3^{+} $0,0,z$
(3) 3^{-} $0,0,z$
(4) 2 $x,x,0$
(5) 2 $x,0,0$
(6) 2 $0,y,0$
(7) $\bar{1}$ $0,0,0$
(8) $\bar{3}^{+}$ $0,0,z;\ 0,0,0$
(9) $\bar{3}^{-}$ $0,0,z;\ 0,0,0$
(10) m $x,\bar{x},z$
(11) m $x,2x,z$
(12) m $2x,x,z$

Generators selected (1); $t(1,0,0)$; $t(0,1,0)$; $t(0,0,1)$; (2); (4); (7)

Positions

Multiplicity, Wyckoff letter, Site symmetry	Coordinates			Reflection conditions

General:

12 j 1

(1) x,y,z (2) $\bar{y},x-y,z$ (3) $\bar{x}+y,\bar{x},z$

(4) $y,x,\bar{z}$ (5) $x-y,\bar{y},\bar{z}$ (6) $\bar{x},\bar{x}+y,\bar{z}$

(7) $\bar{x},\bar{y},\bar{z}$ (8) $y,\bar{x}+y,\bar{z}$ (9) $x-y,x,\bar{z}$

(10) $\bar{y},\bar{x},z$ (11) $\bar{x}+y,y,z$ (12) $x,x-y,z$

no conditions

Special: no extra conditions

6 i . m . $\quad x,\bar{x},z \quad x,2x,z \quad 2\bar{x},\bar{x},z \quad \bar{x},x,\bar{z} \quad 2x,x,\bar{z} \quad \bar{x},2\bar{x},\bar{z}$

6 h . 2 . $\quad x,0,\frac{1}{2} \quad 0,x,\frac{1}{2} \quad \bar{x},\bar{x},\frac{1}{2} \quad \bar{x},0,\frac{1}{2} \quad 0,\bar{x},\frac{1}{2} \quad x,x,\frac{1}{2}$

6 g . 2 . $\quad x,0,0 \quad 0,x,0 \quad \bar{x},\bar{x},0 \quad \bar{x},0,0 \quad 0,\bar{x},0 \quad x,x,0$

3 f . $2/m$. $\quad \frac{1}{2},0,\frac{1}{2} \quad 0,\frac{1}{2},\frac{1}{2} \quad \frac{1}{2},\frac{1}{2},\frac{1}{2}$

3 e . $2/m$. $\quad \frac{1}{2},0,0 \quad 0,\frac{1}{2},0 \quad \frac{1}{2},\frac{1}{2},0$

2 d 3 m . $\quad \frac{1}{3},\frac{2}{3},z \quad \frac{2}{3},\frac{1}{3},\bar{z}$

2 c 3 m . $\quad 0,0,z \quad 0,0,\bar{z}$

1 b $\bar{3}m$. $\quad 0,0,\frac{1}{2}$

1 a $\bar{3}m$. $\quad 0,0,0$

Symmetry of special projections

Along [001] $p6mm$
$\mathbf{a}' = \mathbf{a} \quad \mathbf{b}' = \mathbf{b}$
Origin at $0,0,z$

Along [100] $p2$
$\mathbf{a}' = \frac{1}{2}(\mathbf{a}+2\mathbf{b}) \quad \mathbf{b}' = \mathbf{c}$
Origin at $x,0,0$

Along [210] $p2mm$
$\mathbf{a}' = \frac{1}{2}\mathbf{b} \quad \mathbf{b}' = \mathbf{c}$
Origin at $x,\frac{1}{2}x,0$

Maximal non-isomorphic subgroups

I 　[2] $P3m1$ (156) 　　　1; 2; 3; 10; 11; 12
　　[2] $P321$ (150) 　　　1; 2; 3; 4; 5; 6
　　[2] $P\bar{3}11$ ($P\bar{3}$, 147) 　1; 2; 3; 7; 8; 9
　$\left\{\begin{array}{l}\text{[3] } P12/m1\,(C2/m,\,12) \quad 1;\ 4;\ 7;\ 10 \\ \text{[3] } P12/m1\,(C2/m,\,12) \quad 1;\ 5;\ 7;\ 11 \\ \text{[3] } P12/m1\,(C2/m,\,12) \quad 1;\ 6;\ 7;\ 12\end{array}\right.$

IIa 　none

IIb 　[2] $P\bar{3}c1$ ($\mathbf{c}' = 2\mathbf{c}$) (165); [3] $H\bar{3}m1$ ($\mathbf{a}' = 3\mathbf{a}, \mathbf{b}' = 3\mathbf{b}$) ($P\bar{3}1m$, 162)

Maximal isomorphic subgroups of lowest index

IIc 　[2] $P\bar{3}m1$ ($\mathbf{c}' = 2\mathbf{c}$) (164); [4] $P\bar{3}m1$ ($\mathbf{a}' = 2\mathbf{a}, \mathbf{b}' = 2\mathbf{b}$) (164)

Minimal non-isomorphic supergroups

I 　[2] $P6/mmm$ (191); [2] $P6_3/mmc$ (194)

II 　[3] $H\bar{3}m1$ ($P\bar{3}1m$, 162); [3] $R\bar{3}m$ (obverse) (166); [3] $R\bar{3}m$ (reverse) (166)

$R\bar{3}m$ D_{3d}^{5} $\bar{3}m$ Trigonal

No. 166 $R\bar{3}2/m$

Patterson symmetry $R\bar{3}m$

HEXAGONAL AXES

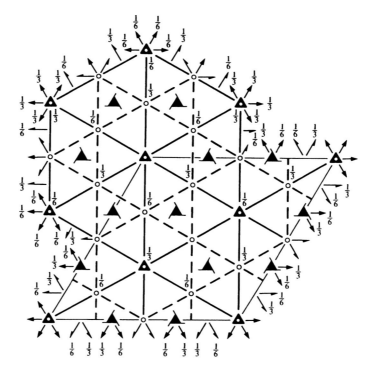

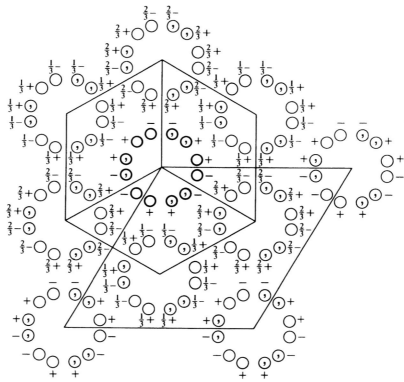

Origin at centre $(\bar{3}m)$

Asymmetric unit $0 \le x \le \frac{2}{3}; \quad 0 \le y \le \frac{2}{3}; \quad 0 \le z \le \frac{1}{6}; \quad x \le 2y; \quad y \le \min(1-x, 2x)$

Vertices $0,0,0 \quad \frac{2}{3},\frac{1}{3},0 \quad \frac{1}{3},\frac{2}{3},0$

$0,0,\frac{1}{6} \quad \frac{2}{3},\frac{1}{3},\frac{1}{6} \quad \frac{1}{3},\frac{2}{3},\frac{1}{6}$

140

Symmetry operations

For $(0,0,0)+$ set

(1) 1	(2) 3^+ $0,0,z$	(3) 3^- $0,0,z$
(4) 2 $x,x,0$	(5) 2 $x,0,0$	(6) 2 $0,y,0$
(7) $\bar{1}$ $0,0,0$	(8) $\bar{3}^+$ $0,0,z$; $0,0,0$	(9) $\bar{3}^-$ $0,0,z$; $0,0,0$
(10) m $x,\bar{x},z$	(11) m $x,2x,z$	(12) m $2x,x,z$

For $(\frac{2}{3},\frac{1}{3},\frac{1}{3})+$ set

(1) $t(\frac{2}{3},\frac{1}{3},\frac{1}{3})$	(2) $3^+(0,0,\frac{1}{3})$ $\frac{1}{3},\frac{1}{3},z$	(3) $3^-(0,0,\frac{1}{3})$ $\frac{1}{3},0,z$
(4) $2(\frac{1}{2},\frac{1}{2},0)$ $x,x-\frac{1}{6},\frac{1}{6}$	(5) $2(\frac{1}{2},0,0)$ $x,\frac{1}{6},\frac{1}{6}$	(6) 2 $\frac{1}{3},y,\frac{1}{6}$
(7) $\bar{1}$ $\frac{1}{3},\frac{1}{6},\frac{1}{6}$	(8) $\bar{3}^+$ $\frac{1}{3},-\frac{1}{3},z$; $\frac{1}{3},-\frac{1}{3},\frac{1}{6}$	(9) $\bar{3}^-$ $\frac{1}{3},\frac{2}{3},z$; $\frac{1}{3},\frac{2}{3},\frac{1}{6}$
(10) $g(\frac{1}{6},-\frac{1}{6},\frac{1}{3})$ $x+\frac{1}{2},\bar{x},z$	(11) $g(\frac{1}{6},\frac{1}{3},\frac{1}{3})$ $x+\frac{1}{4},2x,z$	(12) $g(\frac{2}{3},\frac{1}{3},\frac{1}{3})$ $2x,x,z$

For $(\frac{1}{3},\frac{2}{3},\frac{2}{3})+$ set

(1) $t(\frac{1}{3},\frac{2}{3},\frac{2}{3})$	(2) $3^+(0,0,\frac{2}{3})$ $0,\frac{1}{3},z$	(3) $3^-(0,0,\frac{2}{3})$ $\frac{1}{3},\frac{1}{3},z$
(4) $2(\frac{1}{2},\frac{1}{2},0)$ $x,x+\frac{1}{6},\frac{1}{3}$	(5) 2 $x,\frac{1}{3},\frac{1}{3}$	(6) $2(0,\frac{1}{2},0)$ $\frac{1}{6},y,\frac{1}{3}$
(7) $\bar{1}$ $\frac{1}{6},\frac{1}{3},\frac{1}{3}$	(8) $\bar{3}^+$ $\frac{2}{3},\frac{1}{3},z$; $\frac{2}{3},\frac{1}{3},\frac{1}{3}$	(9) $\bar{3}^-$ $-\frac{1}{3},\frac{1}{3},z$; $-\frac{1}{3},\frac{1}{3},\frac{1}{3}$
(10) $g(-\frac{1}{6},\frac{1}{6},\frac{2}{3})$ $x+\frac{1}{2},\bar{x},z$	(11) $g(\frac{1}{3},\frac{2}{3},\frac{2}{3})$ $x,2x,z$	(12) $g(\frac{1}{3},\frac{1}{6},\frac{2}{3})$ $2x-\frac{1}{2},x,z$

Generators selected   (1); $t(1,0,0)$; $t(0,1,0)$; $t(0,0,1)$; $t(\frac{2}{3},\frac{1}{3},\frac{1}{3})$; (2); (4); (7)

Positions

Multiplicity, Wyckoff letter, Site symmetry	Coordinates $(0,0,0)+$ $(\frac{2}{3},\frac{1}{3},\frac{1}{3})+$ $(\frac{1}{3},\frac{2}{3},\frac{2}{3})+$	Reflection conditions

Reflection conditions

General:

$hkil$: $-h+k+l=3n$
$hki0$: $-h+k=3n$
$hh\overline{2h}l$: $l=3n$
$h\bar{h}0l$: $h+l=3n$
$000l$: $l=3n$
$h\bar{h}00$: $h=3n$

Special: no extra conditions

36	i	1			

(1) x,y,z (2) $\bar{y},x-y,z$ (3) $\bar{x}+y,\bar{x},z$
(4) $y,x,\bar{z}$ (5) $x-y,\bar{y},\bar{z}$ (6) $\bar{x},\bar{x}+y,\bar{z}$
(7) $\bar{x},\bar{y},\bar{z}$ (8) $y,\bar{x}+y,\bar{z}$ (9) $x-y,x,\bar{z}$
(10) $\bar{y},\bar{x},z$ (11) $\bar{x}+y,y,z$ (12) $x,x-y,z$

18	h	$.m$	$x,\bar{x},z$	$x,2x,z$	$2\bar{x},\bar{x},z$	$\bar{x},x,\bar{z}$	$2x,x,\bar{z}$	$\bar{x},2\bar{x},\bar{z}$
18	g	$.2$	$x,0,\frac{1}{2}$	$0,x,\frac{1}{2}$	$\bar{x},\bar{x},\frac{1}{2}$	$\bar{x},0,\frac{1}{2}$	$0,\bar{x},\frac{1}{2}$	$x,x,\frac{1}{2}$
18	f	$.2$	$x,0,0$	$0,x,0$	$\bar{x},\bar{x},0$	$\bar{x},0,0$	$0,\bar{x},0$	$x,x,0$
9	e	$.2/m$	$\frac{1}{2},0,0$	$0,\frac{1}{2},0$	$\frac{1}{2},\frac{1}{2},0$			
9	d	$.2/m$	$\frac{1}{2},0,\frac{1}{2}$	$0,\frac{1}{2},\frac{1}{2}$	$\frac{1}{2},\frac{1}{2},\frac{1}{2}$			
6	c	$3m$	$0,0,z$	$0,0,\bar{z}$				
3	b	$\bar{3}m$	$0,0,\frac{1}{2}$					
3	a	$\bar{3}m$	$0,0,0$					

Symmetry of special projections

Along [001] $p6mm$	Along [100] $p2$	Along [210] $p2mm$
$\mathbf{a}'=\frac{1}{3}(2\mathbf{a}+\mathbf{b})$ $\mathbf{b}'=\frac{1}{3}(-\mathbf{a}+\mathbf{b})$	$\mathbf{a}'=\frac{1}{2}(\mathbf{a}+2\mathbf{b})$ $\mathbf{b}'=\frac{1}{3}(-\mathbf{a}-2\mathbf{b}+\mathbf{c})$	$\mathbf{a}'=\frac{1}{2}\mathbf{b}$ $\mathbf{b}'=\frac{1}{3}\mathbf{c}$
Origin at $0,0,z$	Origin at $x,0,0$	Origin at $x,\frac{1}{2}x,0$

HEXAGONAL AXES

Maximal non-isomorphic subgroups

I [2] $R3m$ (160) (1; 2; 3; 10; 11; 12)+
  [2] $R32$ (155) (1; 2; 3; 4; 5; 6)+
  [2] $R\bar{3}1$ ($R\bar{3}$, 148) (1; 2; 3; 7; 8; 9)+
  ⎧ [3] $R12/m$ ($C2/m$, 12) (1; 4; 7; 10)+
  ⎨ [3] $R12/m$ ($C2/m$, 12) (1; 5; 7; 11)+
  ⎩ [3] $R12/m$ ($C2/m$, 12) (1; 6; 7; 12)+

IIa ⎧ [3] $P\bar{3}m1$ (164) 1; 2; 3; 4; 5; 6; 7; 8; 9; 10; 11; 12
  ⎨ [3] $P\bar{3}m1$ (164) 1; 2; 3; 10; 11; 12; (4; 5; 6; 7; 8; 9) + $(\frac{2}{3}, \frac{1}{3}, \frac{1}{3})$
  ⎩ [3] $P\bar{3}m1$ (164) 1; 2; 3; 10; 11; 12; (4; 5; 6; 7; 8; 9) + $(\frac{1}{3}, \frac{2}{3}, \frac{2}{3})$

IIb [2] $R\bar{3}c$ ($\mathbf{a}' = -\mathbf{a}, \mathbf{b}' = -\mathbf{b}, \mathbf{c}' = 2\mathbf{c}$) (167)

Maximal isomorphic subgroups of lowest index

IIc [2] $R\bar{3}m$ ($\mathbf{a}' = -\mathbf{a}, \mathbf{b}' = -\mathbf{b}, \mathbf{c}' = 2\mathbf{c}$) (166); [4] $R\bar{3}m$ ($\mathbf{a}' = -2\mathbf{a}, \mathbf{b}' = -2\mathbf{b}$) (166)

Minimal non-isomorphic supergroups

I [4] $Pm\bar{3}m$ (221); [4] $Pn\bar{3}m$ (224); [4] $Fm\bar{3}m$ (225); [4] $Fd\bar{3}m$ (227); [4] $Im\bar{3}m$ (229)

II [3] $P\bar{3}1m$ ($\mathbf{a}' = \frac{1}{3}(2\mathbf{a}+\mathbf{b}), \mathbf{b}' = \frac{1}{3}(-\mathbf{a}+\mathbf{b}), \mathbf{c}' = \frac{1}{3}\mathbf{c}$) (162)

RHOMBOHEDRAL AXES

Maximal non-isomorphic subgroups

I [2] $R3m$ (160) 1; 2; 3; 10; 11; 12
  [2] $R32$ (155) 1; 2; 3; 4; 5; 6
  [2] $R\bar{3}1$ ($R\bar{3}$, 148) 1; 2; 3; 7; 8; 9
  ⎧ [3] $R12/m$ ($C2/m$, 12) 1; 4; 7; 10
  ⎨ [3] $R12/m$ ($C2/m$, 12) 1; 5; 7; 11
  ⎩ [3] $R12/m$ ($C2/m$, 12) 1; 6; 7; 12

IIa none

IIb [2] $F\bar{3}c$ ($\mathbf{a}' = 2\mathbf{a}, \mathbf{b}' = 2\mathbf{b}, \mathbf{c}' = 2\mathbf{c}$) ($R\bar{3}c$, 167); [3] $P\bar{3}m1$ ($\mathbf{a}' = \mathbf{a}-\mathbf{b}, \mathbf{b}' = \mathbf{b}-\mathbf{c}, \mathbf{c}' = \mathbf{a}+\mathbf{b}+\mathbf{c}$) (164)

Maximal isomorphic subgroups of lowest index

IIc [2] $R\bar{3}m$ ($\mathbf{a}' = \mathbf{b}+\mathbf{c}, \mathbf{b}' = \mathbf{a}+\mathbf{c}, \mathbf{c}' = \mathbf{a}+\mathbf{b}$) (166); [4] $R\bar{3}m$ ($\mathbf{a}' = -\mathbf{a}+\mathbf{b}+\mathbf{c}, \mathbf{b}' = \mathbf{a}-\mathbf{b}+\mathbf{c}, \mathbf{c}' = \mathbf{a}+\mathbf{b}-\mathbf{c}$) (166)

Minimal non-isomorphic supergroups

I [4] $Pm\bar{3}m$ (221); [4] $Pn\bar{3}m$ (224); [4] $Fm\bar{3}m$ (225); [4] $Fd\bar{3}m$ (227); [4] $Im\bar{3}m$ (229)

II [3] $P\bar{3}1m$ ($\mathbf{a}' = \frac{1}{3}(2\mathbf{a}-\mathbf{b}-\mathbf{c}), \mathbf{b}' = \frac{1}{3}(-\mathbf{a}+2\mathbf{b}-\mathbf{c}), \mathbf{c}' = \frac{1}{3}(\mathbf{a}+\mathbf{b}+\mathbf{c})$) (162)

RHOMBOHEDRAL AXES
(For drawings see hexagonal axes)

Origin at centre $(\bar{3}m)$

Asymmetric unit $0 \le x \le 1;$ $0 \le y \le 1;$ $0 \le z \le \frac{1}{2};$ $y \le x;$ $z \le \min(y, 1-x)$

    Vertices $0,0,0$ $1,0,0$ $1,1,0$ $\frac{1}{2},\frac{1}{2},\frac{1}{2}$

Symmetry operations

(1) 1	(2) 3^+ x,x,x	(3) 3^- x,x,x
(4) 2 $\bar{x},0,x$	(5) 2 $x,\bar{x},0$	(6) 2 $0,y,\bar{y}$
(7) $\bar{1}$ $0,0,0$	(8) $\bar{3}^+$ x,x,x; $0,0,0$	(9) $\bar{3}^-$ x,x,x; $0,0,0$
(10) m x,y,x	(11) m x,x,z	(12) m x,y,y

Generators selected   (1); $t(1,0,0)$; $t(0,1,0)$; $t(0,0,1)$; (2); (4); (7)

Positions

Multiplicity, Wyckoff letter, Site symmetry			Coordinates					Reflection conditions

General:

12	i	1	(1) x,y,z	(2) z,x,y	(3) y,z,x			no conditions
			(4) $\bar{z},\bar{y},\bar{x}$	(5) $\bar{y},\bar{x},\bar{z}$	(6) $\bar{x},\bar{z},\bar{y}$			
			(7) $\bar{x},\bar{y},\bar{z}$	(8) $\bar{z},\bar{x},\bar{y}$	(9) $\bar{y},\bar{z},\bar{x}$			
			(10) z,y,x	(11) y,x,z	(12) x,z,y			

Special: no extra conditions

6	h	$.m$	x,x,z	z,x,x	x,z,x	$\bar{z},\bar{x},\bar{x}$	$\bar{x},\bar{x},\bar{z}$	$\bar{x},\bar{z},\bar{x}$
6	g	$.2$	$x,\bar{x},\frac{1}{2}$	$\frac{1}{2},x,\bar{x}$	$\bar{x},\frac{1}{2},x$	$\bar{x},x,\frac{1}{2}$	$\frac{1}{2},\bar{x},x$	$x,\frac{1}{2},\bar{x}$
6	f	$.2$	$x,\bar{x},0$	$0,x,\bar{x}$	$\bar{x},0,x$	$\bar{x},x,0$	$0,\bar{x},x$	$x,0,\bar{x}$
3	e	$.2/m$	$0,\frac{1}{2},\frac{1}{2}$	$\frac{1}{2},0,\frac{1}{2}$	$\frac{1}{2},\frac{1}{2},0$			
3	d	$.2/m$	$\frac{1}{2},0,0$	$0,\frac{1}{2},0$	$0,0,\frac{1}{2}$			
2	c	$3m$	x,x,x	$\bar{x},\bar{x},\bar{x}$				
1	b	$\bar{3}m$	$\frac{1}{2},\frac{1}{2},\frac{1}{2}$					
1	a	$\bar{3}m$	$0,0,0$					

Symmetry of special projections

Along $[111]$ $p6mm$
$\mathbf{a}' = \frac{1}{3}(2\mathbf{a} - \mathbf{b} - \mathbf{c})$ $\mathbf{b}' = \frac{1}{3}(-\mathbf{a} + 2\mathbf{b} - \mathbf{c})$
Origin at x,x,x

Along $[1\bar{1}0]$ $p2$
$\mathbf{a}' = \frac{1}{2}(\mathbf{a} + \mathbf{b} - 2\mathbf{c})$ $\mathbf{b}' = \mathbf{c}$
Origin at $x,\bar{x},0$

Along $[2\bar{1}\bar{1}]$ $p2mm$
$\mathbf{a}' = \frac{1}{2}(\mathbf{b} - \mathbf{c})$ $\mathbf{b}' = \frac{1}{3}(\mathbf{a} + \mathbf{b} + \mathbf{c})$
Origin at $2x,\bar{x},\bar{x}$

(*Continued on preceding page*)

$P6_3/mmc$ D_{6h}^4 $6/mmm$ Hexagonal

No. 194 $P\ 6_3/m\ 2/m\ 2/c$ Patterson symmetry $P6/mmm$

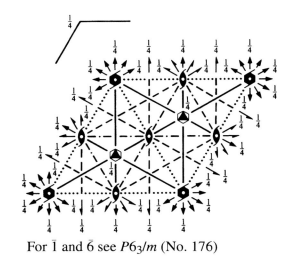 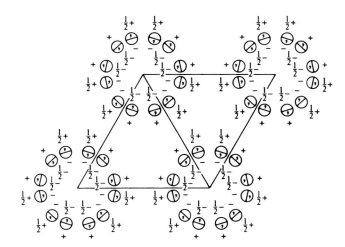

For $\bar{1}$ and $\bar{6}$ see $P6_3/m$ (No. 176)

Origin at centre $(\bar{3}m1)$ at $\bar{3}2/mc$

Asymmetric unit $0 \le x \le \frac{2}{3}$; $0 \le y \le \frac{2}{3}$; $0 \le z \le \frac{1}{4}$; $x \le 2y$; $y \le \min(1-x, 2x)$

Vertices $0,0,0$ $\frac{2}{3},\frac{1}{3},0$ $\frac{1}{3},\frac{2}{3},0$

$0,0,\frac{1}{4}$ $\frac{2}{3},\frac{1}{3},\frac{1}{4}$ $\frac{1}{3},\frac{2}{3},\frac{1}{4}$

Symmetry operations

(1) 1
(2) 3^+ $0,0,z$
(3) 3^- $0,0,z$
(4) $2(0,0,\frac{1}{2})$ $0,0,z$
(5) $6^-(0,0,\frac{1}{2})$ $0,0,z$
(6) $6^+(0,0,\frac{1}{2})$ $0,0,z$
(7) 2 $x,x,0$
(8) 2 $x,0,0$
(9) 2 $0,y,0$
(10) 2 $x,\bar{x},\frac{1}{4}$
(11) 2 $x,2x,\frac{1}{4}$
(12) 2 $2x,x,\frac{1}{4}$
(13) $\bar{1}$ $0,0,0$
(14) $\bar{3}^+$ $0,0,z$; $0,0,0$
(15) $\bar{3}^-$ $0,0,z$; $0,0,0$
(16) m $x,y,\frac{1}{4}$
(17) $\bar{6}^-$ $0,0,z$; $0,0,\frac{1}{4}$
(18) $\bar{6}^+$ $0,0,z$; $0,0,\frac{1}{4}$
(19) m $x,\bar{x},z$
(20) m $x,2x,z$
(21) m $2x,x,z$
(22) c x,x,z
(23) c $x,0,z$
(24) c $0,y,z$

Maximal non-isomorphic subgroups

I [2] $P\bar{6}2c$ (190) 1; 2; 3; 7; 8; 9; 16; 17; 18; 22; 23; 24
  [2] $P\bar{6}m2$ (187) 1; 2; 3; 10; 11; 12; 16; 17; 18; 19; 20; 21
  [2] $P6_3mc$ (186) 1; 2; 3; 4; 5; 6; 19; 20; 21; 22; 23; 24
  [2] $P6_322$ (182) 1; 2; 3; 4; 5; 6; 7; 8; 9; 10; 11; 12
  [2] $P6_3/m11$ ($P6_3/m$, 176) 1; 2; 3; 4; 5; 6; 13; 14; 15; 16; 17; 18
  [2] $P\bar{3}m1$ (164) 1; 2; 3; 7; 8; 9; 13; 14; 15; 19; 20; 21
  [2] $P\bar{3}1c$ (163) 1; 2; 3; 10; 11; 12; 13; 14; 15; 22; 23; 24
  ⎰ [3] $Pmmc$ ($Cmcm$, 63) 1; 4; 7; 10; 13; 16; 19; 22
  ⎨ [3] $Pmmc$ ($Cmcm$, 63) 1; 4; 8; 11; 13; 16; 20; 23
  ⎱ [3] $Pmmc$ ($Cmcm$, 63) 1; 4; 9; 12; 13; 16; 21; 24

IIa none
IIb [3] $H6_3/mmc$ ($\mathbf{a}' = 3\mathbf{a}, \mathbf{b}' = 3\mathbf{b}$) ($P6_3/mcm$, 193)

Maximal isomorphic subgroups of lowest index

IIc [3] $P6_3/mmc$ ($\mathbf{c}' = 3\mathbf{c}$) (194); [4] $P6_3/mmc$ ($\mathbf{a}' = 2\mathbf{a}, \mathbf{b}' = 2\mathbf{b}$) (194)

Minimal non-isomorphic supergroups

I none
II [3] $H6_3/mmc$ ($P6_3/mcm$, 193); [2] $P6/mmm$ ($\mathbf{c}' = \frac{1}{2}\mathbf{c}$) (191)

Generators selected   (1); $t(1,0,0)$; $t(0,1,0)$; $t(0,0,1)$; (2); (4); (7); (13)

Positions

Multiplicity, Wyckoff letter, Site symmetry	Coordinates	Reflection conditions

Coordinates / General:

24   l 1

(1) x,y,z (2) $\bar{y},x-y,z$ (3) $\bar{x}+y,\bar{x},z$
(4) $\bar{x},\bar{y},z+\frac{1}{2}$ (5) $y,\bar{x}+y,z+\frac{1}{2}$ (6) $x-y,x,z+\frac{1}{2}$
(7) $y,x,\bar{z}$ (8) $x-y,\bar{y},\bar{z}$ (9) $\bar{x},\bar{x}+y,\bar{z}$
(10) $\bar{y},\bar{x},\bar{z}+\frac{1}{2}$ (11) $\bar{x}+y,y,\bar{z}+\frac{1}{2}$ (12) $x,x-y,\bar{z}+\frac{1}{2}$
(13) $\bar{x},\bar{y},\bar{z}$ (14) $y,\bar{x}+y,\bar{z}$ (15) $x-y,x,\bar{z}$
(16) $x,y,\bar{z}+\frac{1}{2}$ (17) $\bar{y},x-y,\bar{z}+\frac{1}{2}$ (18) $\bar{x}+y,\bar{x},\bar{z}+\frac{1}{2}$
(19) $\bar{y},\bar{x},z$ (20) $\bar{x}+y,y,z$ (21) $x,x-y,z$
(22) $y,x,z+\frac{1}{2}$ (23) $x-y,\bar{y},z+\frac{1}{2}$ (24) $\bar{x},\bar{x}+y,z+\frac{1}{2}$

General:
$hh\overline{2h}l$: $l=2n$
$000l$: $l=2n$

Special: as above, plus

12   k $.m.$

$x,2x,z$ $2\bar{x},\bar{x},z$ $x,\bar{x},z$ $\bar{x},2\bar{x},z+\frac{1}{2}$
$2x,x,z+\frac{1}{2}$ $\bar{x},x,z+\frac{1}{2}$ $2x,x,\bar{z}$ $\bar{x},2\bar{x},\bar{z}$
$\bar{x},x,\bar{z}$ $2\bar{x},\bar{x},\bar{z}+\frac{1}{2}$ $x,2x,\bar{z}+\frac{1}{2}$ $x,\bar{x},\bar{z}+\frac{1}{2}$

no extra conditions

12   j $m..$

$x,y,\frac{1}{4}$ $\bar{y},x-y,\frac{1}{4}$ $\bar{x}+y,\bar{x},\frac{1}{4}$ $\bar{x},\bar{y},\frac{3}{4}$ $y,\bar{x}+y,\frac{3}{4}$ $x-y,x,\frac{3}{4}$
$y,x,\frac{3}{4}$ $x-y,\bar{y},\frac{3}{4}$ $\bar{x},\bar{x}+y,\frac{3}{4}$ $\bar{y},\bar{x},\frac{1}{4}$ $\bar{x}+y,y,\frac{1}{4}$ $x,x-y,\frac{1}{4}$

no extra conditions

12   i $.2.$

$x,0,0$ $0,x,0$ $\bar{x},\bar{x},0$ $\bar{x},0,\frac{1}{2}$ $0,\bar{x},\frac{1}{2}$ $x,x,\frac{1}{2}$
$\bar{x},0,0$ $0,\bar{x},0$ $x,x,0$ $x,0,\frac{1}{2}$ $0,x,\frac{1}{2}$ $\bar{x},\bar{x},\frac{1}{2}$

$hkil$: $l=2n$

6   h $mm2$

$x,2x,\frac{1}{4}$ $2\bar{x},\bar{x},\frac{1}{4}$ $x,\bar{x},\frac{1}{4}$ $\bar{x},2\bar{x},\frac{3}{4}$ $2x,x,\frac{3}{4}$ $\bar{x},x,\frac{3}{4}$

no extra conditions

6   g $.2/m.$

$\frac{1}{2},0,0$ $0,\frac{1}{2},0$ $\frac{1}{2},\frac{1}{2},0$ $\frac{1}{2},0,\frac{1}{2}$ $0,\frac{1}{2},\frac{1}{2}$ $\frac{1}{2},\frac{1}{2},\frac{1}{2}$

$hkil$: $l=2n$

4   f $3m.$

$\frac{1}{3},\frac{2}{3},z$ $\frac{2}{3},\frac{1}{3},z+\frac{1}{2}$ $\frac{2}{3},\frac{1}{3},\bar{z}$ $\frac{1}{3},\frac{2}{3},\bar{z}+\frac{1}{2}$

$hkil$: $l=2n$
or $h-k=3n+1$
or $h-k=3n+2$

4   e $3m.$

$0,0,z$ $0,0,z+\frac{1}{2}$ $0,0,\bar{z}$ $0,0,\bar{z}+\frac{1}{2}$

$hkil$: $l=2n$

2   d $\bar{6}m2$

$\frac{1}{3},\frac{2}{3},\frac{3}{4}$ $\frac{2}{3},\frac{1}{3},\frac{1}{4}$

$hkil$: $l=2n$
or $h-k=3n+1$
or $h-k=3n+2$

2   c $\bar{6}m2$

$\frac{1}{3},\frac{2}{3},\frac{1}{4}$ $\frac{2}{3},\frac{1}{3},\frac{3}{4}$

2   b $\bar{6}m2$

$0,0,\frac{1}{4}$ $0,0,\frac{3}{4}$

$hkil$: $l=2n$

2   a $\bar{3}m.$

$0,0,0$ $0,0,\frac{1}{2}$

$hkil$: $l=2n$

Symmetry of special projections

Along [001] $p6mm$
$\mathbf{a'}=\mathbf{a}$ $\mathbf{b'}=\mathbf{b}$
Origin at $0,0,z$

Along [100] $p2gm$
$\mathbf{a'}=\frac{1}{2}(\mathbf{a}+2\mathbf{b})$ $\mathbf{b'}=\mathbf{c}$
Origin at $x,0,0$

Along [210] $p2mm$
$\mathbf{a'}=\frac{1}{2}\mathbf{b}$ $\mathbf{b'}=\frac{1}{2}\mathbf{c}$
Origin at $x,\frac{1}{2}x,0$

(*Continued on preceding page*)

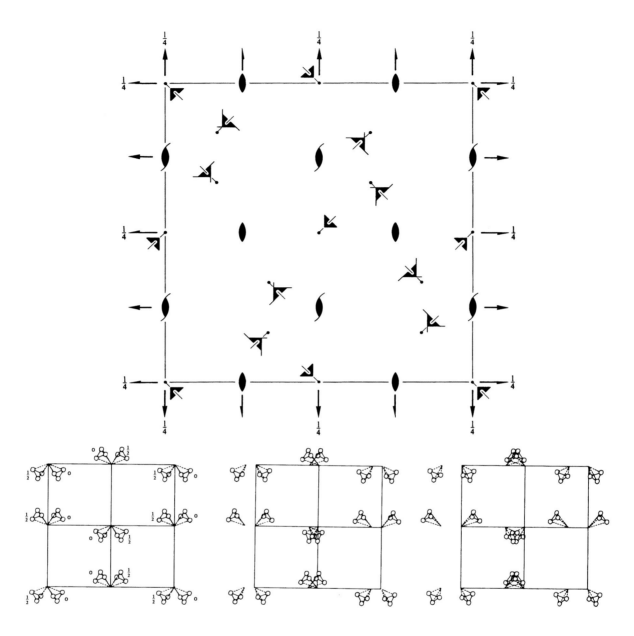

Origin on 3[111] at midpoint of three non-intersecting pairs of parallel 2 axes and of three non-intersecting pairs of parallel 2_1 axes

Asymmetric unit $0 \le x \le \frac{1}{2}; \quad 0 \le y \le \frac{1}{2}; \quad 0 \le z \le \frac{1}{2}; \quad z \le \min(x,y)$

Vertices $0,0,0 \quad \frac{1}{2},0,0 \quad \frac{1}{2},\frac{1}{2},0 \quad 0,\frac{1}{2},0 \quad \frac{1}{2},\frac{1}{2},\frac{1}{2}$

Symmetry operations

For $(0,0,0)+$ set

(1) 1

(2) $2(0,0,\frac{1}{2})$ $\frac{1}{4},0,z$

(3) $2(0,\frac{1}{2},0)$ $0,y,\frac{1}{4}$

(4) $2(\frac{1}{2},0,0)$ $x,\frac{1}{4},0$

(5) 3^+ x,x,x

(6) 3^+ $\bar{x}+\frac{1}{2},x,\bar{x}$

(7) 3^+ $x+\frac{1}{2},\bar{x}-\frac{1}{2},\bar{x}$

(8) 3^+ $\bar{x},\bar{x}+\frac{1}{2},x$

(9) 3^- x,x,x

(10) $3^-\left(-\frac{1}{3},\frac{1}{3},\frac{1}{3}\right)$ $x+\frac{1}{6},\bar{x}+\frac{1}{6},\bar{x}$

(11) $3^-\left(\frac{1}{3},\frac{1}{3},-\frac{1}{3}\right)$ $\bar{x}+\frac{1}{3},\bar{x}+\frac{1}{6},x$

(12) $3^-\left(\frac{1}{3},-\frac{1}{3},\frac{1}{3}\right)$ $\bar{x}-\frac{1}{6},x+\frac{1}{3},\bar{x}$

For $\left(\frac{1}{2},\frac{1}{2},\frac{1}{2}\right)+$ set

(1) $t\left(\frac{1}{2},\frac{1}{2},\frac{1}{2}\right)$

(2) 2 $0,\frac{1}{4},z$

(3) 2 $\frac{1}{4},y,0$

(4) 2 $x,0,\frac{1}{4}$

(5) $3^+\left(\frac{1}{2},\frac{1}{2},\frac{1}{2}\right)$ x,x,x

(6) $3^+\left(\frac{1}{6},-\frac{1}{6},\frac{1}{6}\right)$ $\bar{x}-\frac{1}{6},x+\frac{1}{3},\bar{x}$

(7) $3^+\left(-\frac{1}{6},\frac{1}{6},\frac{1}{6}\right)$ $x+\frac{1}{6},\bar{x}+\frac{1}{6},\bar{x}$

(8) $3^+\left(\frac{1}{6},\frac{1}{6},-\frac{1}{6}\right)$ $\bar{x}+\frac{1}{3},\bar{x}+\frac{1}{6},x$

(9) $3^-\left(\frac{1}{2},\frac{1}{2},\frac{1}{2}\right)$ x,x,x

(10) $3^-\left(\frac{1}{6},-\frac{1}{6},-\frac{1}{6}\right)$ $x+\frac{1}{6},\bar{x}+\frac{1}{6},\bar{x}$

(11) $3^-\left(-\frac{1}{6},-\frac{1}{6},\frac{1}{6}\right)$ $\bar{x}+\frac{1}{3},\bar{x}+\frac{1}{6},x$

(12) $3^-\left(-\frac{1}{6},\frac{1}{6},-\frac{1}{6}\right)$ $\bar{x}-\frac{1}{6},x+\frac{1}{3},\bar{x}$

Generators selected   (1); $t(1,0,0)$; $t(0,1,0)$; $t(0,0,1)$; $t(\frac{1}{2},\frac{1}{2},\frac{1}{2})$; (2); (3); (5)

Positions

Multiplicity, Wyckoff letter, Site symmetry	Coordinates $(0,0,0)+$ $(\frac{1}{2},\frac{1}{2},\frac{1}{2})+$			Reflection conditions h,k,l cyclically permutable General:

24   c 1

    (1) x,y,z (2) $\bar{x}+\frac{1}{2},\bar{y},z+\frac{1}{2}$ (3) $\bar{x},y+\frac{1}{2},\bar{z}+\frac{1}{2}$ (4) $x+\frac{1}{2},\bar{y}+\frac{1}{2},\bar{z}$ $hkl: h+k+l=2n$

    (5) z,x,y (6) $z+\frac{1}{2},\bar{x}+\frac{1}{2},\bar{y}$ (7) $\bar{z}+\frac{1}{2},\bar{x},y+\frac{1}{2}$ (8) $\bar{z},x+\frac{1}{2},\bar{y}+\frac{1}{2}$ $0kl: k+l=2n$

    (9) y,z,x (10) $\bar{y},z+\frac{1}{2},\bar{x}+\frac{1}{2}$ (11) $y+\frac{1}{2},\bar{z}+\frac{1}{2},\bar{x}$ (12) $\bar{y}+\frac{1}{2},\bar{z},x+\frac{1}{2}$ $hhl: l=2n$

            $h00: h=2n$

    Special: no extra conditions

12   b $2..$ $x,0,\frac{1}{4}$ $\bar{x}+\frac{1}{2},0,\frac{3}{4}$ $\frac{1}{4},x,0$ $\frac{3}{4},\bar{x}+\frac{1}{2},0$ $0,\frac{1}{4},x$ $0,\frac{3}{4},\bar{x}+\frac{1}{2}$

8   a $.3.$ x,x,x $\bar{x}+\frac{1}{2},\bar{x},x+\frac{1}{2}$ $\bar{x},x+\frac{1}{2},\bar{x}+\frac{1}{2}$ $x+\frac{1}{2},\bar{x}+\frac{1}{2},\bar{x}$

Symmetry of special projections

Along $[001]$ $c2mm$ Along $[111]$ $p3$ Along $[110]$ $p1m1$

$\mathbf{a}'=\mathbf{a}$ $\mathbf{b}'=\mathbf{b}$ $\mathbf{a}'=\frac{1}{3}(2\mathbf{a}-\mathbf{b}-\mathbf{c})$ $\mathbf{b}'=\frac{1}{3}(-\mathbf{a}+2\mathbf{b}-\mathbf{c})$ $\mathbf{a}'=\frac{1}{2}(-\mathbf{a}+\mathbf{b})$ $\mathbf{b}'=\frac{1}{2}\mathbf{c}$

Origin at $\frac{1}{4},0,z$ Origin at x,x,x Origin at $x,x+\frac{1}{4},0$

Maximal non-isomorphic subgroups

I $[3]\,I2_11\,(I2_12_12_1,24)$ $(1;\ 2;\ 3;\ 4)+$

      $\Big\{$ $[4]\,I13\,(R3,146)$ $(1;\ 5;\ 9)+$

         $[4]\,I13\,(R3,146)$ $(1;\ 6;\ 12)+$

         $[4]\,I13\,(R3,146)$ $(1;\ 7;\ 10)+$

         $[4]\,I13\,(R3,146)$ $(1;\ 8;\ 11)+$

IIa $[2]\,P2_13\,(198)$ $1;\ 2;\ 3;\ 4;\ 5;\ 6;\ 7;\ 8;\ 9;\ 10;\ 11;\ 12$

IIb   none

Maximal isomorphic subgroups of lowest index

IIc $[27]\,I2_13\,(\mathbf{a}'=3\mathbf{a},\mathbf{b}'=3\mathbf{b},\mathbf{c}'=3\mathbf{c})\,(199)$

Minimal non-isomorphic supergroups

I $[2]\,Ia\bar{3}\,(206);\ [2]\,I4_132\,(214);\ [2]\,I\bar{4}3d\,(220)$

II $[4]\,P23\,(\mathbf{a}'=\frac{1}{2}\mathbf{a},\mathbf{b}'=\frac{1}{2}\mathbf{b},\mathbf{c}'=\frac{1}{2}\mathbf{c})\,(195)$

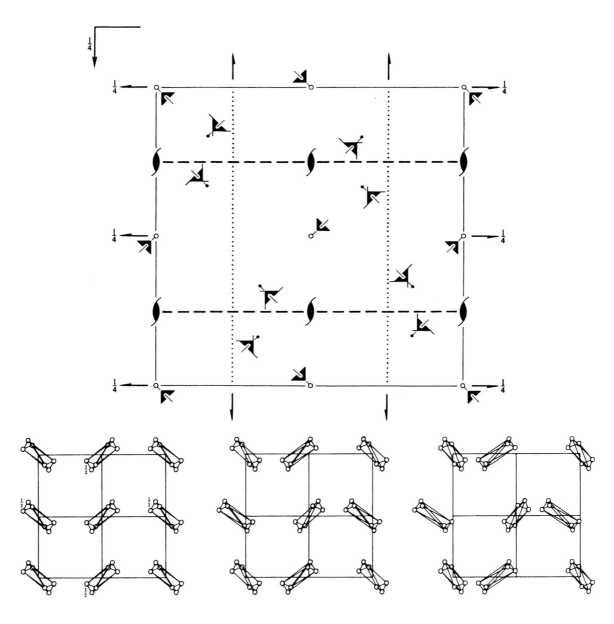

Origin at centre $(\bar{3})$

Asymmetric unit $0 \le x \le \frac{1}{2}$; $0 \le y \le \frac{1}{2}$; $0 \le z \le \frac{1}{2}$; $z \le \min(x,y)$

Vertices $0,0,0$ $\frac{1}{2},0,0$ $\frac{1}{2},\frac{1}{2},0$ $0,\frac{1}{2},0$ $\frac{1}{2},\frac{1}{2},\frac{1}{2}$

Symmetry operations

(1) 1

(2) $2(0,0,\frac{1}{2})$ $\frac{1}{4},0,z$

(3) $2(0,\frac{1}{2},0)$ $0,y,\frac{1}{4}$

(4) $2(\frac{1}{2},0,0)$ $x,\frac{1}{4},0$

(5) 3^+ x,x,x

(6) 3^+ $\bar{x}+\frac{1}{2},x,\bar{x}$

(7) 3^+ $x+\frac{1}{2},\bar{x}-\frac{1}{2},\bar{x}$

(8) 3^+ $\bar{x},\bar{x}+\frac{1}{2},x$

(9) 3^- x,x,x

(10) $3^-\left(-\frac{1}{3},\frac{1}{3},\frac{1}{3}\right)$ $x+\frac{1}{6},\bar{x}+\frac{1}{6},\bar{x}$

(11) $3^-\left(\frac{1}{3},\frac{1}{3},-\frac{1}{3}\right)$ $\bar{x}+\frac{1}{3},\bar{x}+\frac{1}{6},x$

(12) $3^-\left(\frac{1}{3},-\frac{1}{3},\frac{1}{3}\right)$ $\bar{x}-\frac{1}{6},x+\frac{1}{3},\bar{x}$

(13) $\bar{1}$ $0,0,0$

(14) a $x,y,\frac{1}{4}$

(15) c $x,\frac{1}{4},z$

(16) b $\frac{1}{4},y,z$

(17) $\bar{3}^+$ x,x,x; $0,0,0$

(18) $\bar{3}^+$ $\bar{x}-\frac{1}{2},x+1,\bar{x}$; $0,\frac{1}{2},\frac{1}{2}$

(19) $\bar{3}^+$ $x+\frac{1}{2},\bar{x}+\frac{1}{2},\bar{x}$; $\frac{1}{2},\frac{1}{2},0$

(20) $\bar{3}^+$ $\bar{x}+1,\bar{x}+\frac{1}{2},x$; $\frac{1}{2},0,\frac{1}{2}$

(21) $\bar{3}^-$ x,x,x; $0,0,0$

(22) $\bar{3}^-$ $x+\frac{1}{2},\bar{x}-\frac{1}{2},\bar{x}$; $0,0,\frac{1}{2}$

(23) $\bar{3}^-$ $\bar{x},\bar{x}+\frac{1}{2},x$; $0,\frac{1}{2},0$

(24) $\bar{3}^-$ $\bar{x}+\frac{1}{2},x,\bar{x}$; $\frac{1}{2},0,0$

Generators selected   (1); $t(1,0,0)$; $t(0,1,0)$; $t(0,0,1)$; (2); (3); (5); (13)

Positions

Multiplicity, Wyckoff letter, Site symmetry	Coordinates				Reflection conditions

h,k,l cyclically permutable

General:

24   **d**   1

(1) x,y,z (2) $\bar{x}+\frac{1}{2},\bar{y},z+\frac{1}{2}$ (3) $\bar{x},y+\frac{1}{2},\bar{z}+\frac{1}{2}$ (4) $x+\frac{1}{2},\bar{y}+\frac{1}{2},\bar{z}$ $0kl: \; k=2n$

(5) z,x,y (6) $z+\frac{1}{2},\bar{x}+\frac{1}{2},\bar{y}$ (7) $\bar{z}+\frac{1}{2},\bar{x},y+\frac{1}{2}$ (8) $\bar{z},x+\frac{1}{2},\bar{y}+\frac{1}{2}$ $h00: \; h=2n$

(9) y,z,x (10) $\bar{y},z+\frac{1}{2},\bar{x}+\frac{1}{2}$ (11) $y+\frac{1}{2},\bar{z}+\frac{1}{2},\bar{x}$ (12) $\bar{y}+\frac{1}{2},\bar{z},x+\frac{1}{2}$

(13) $\bar{x},\bar{y},\bar{z}$ (14) $x+\frac{1}{2},y,\bar{z}+\frac{1}{2}$ (15) $x,\bar{y}+\frac{1}{2},z+\frac{1}{2}$ (16) $\bar{x}+\frac{1}{2},y+\frac{1}{2},z$

(17) $\bar{z},\bar{x},\bar{y}$ (18) $\bar{z}+\frac{1}{2},x+\frac{1}{2},y$ (19) $z+\frac{1}{2},x,\bar{y}+\frac{1}{2}$ (20) $z,\bar{x}+\frac{1}{2},y+\frac{1}{2}$

(21) $\bar{y},\bar{z},\bar{x}$ (22) $y,\bar{z}+\frac{1}{2},x+\frac{1}{2}$ (23) $\bar{y}+\frac{1}{2},z+\frac{1}{2},x$ (24) $y+\frac{1}{2},z,\bar{x}+\frac{1}{2}$

Special: as above, plus

8   **c**   .3.

x,x,x $\bar{x}+\frac{1}{2},\bar{x},x+\frac{1}{2}$ $\bar{x},x+\frac{1}{2},\bar{x}+\frac{1}{2}$ $x+\frac{1}{2},\bar{x}+\frac{1}{2},\bar{x}$ no extra conditions

$\bar{x},\bar{x},\bar{x}$ $x+\frac{1}{2},x,\bar{x}+\frac{1}{2}$ $x,\bar{x}+\frac{1}{2},x+\frac{1}{2}$ $\bar{x}+\frac{1}{2},x+\frac{1}{2},x$

4   **b**   .$\bar{3}$.

$\frac{1}{2},\frac{1}{2},\frac{1}{2}$ $0,\frac{1}{2},0$ $\frac{1}{2},0,0$ $0,0,\frac{1}{2}$ $hkl: \; h+k,h+l,k+l=2n$

4   **a**   .$\bar{3}$.

$0,0,0$ $\frac{1}{2},0,\frac{1}{2}$ $0,\frac{1}{2},\frac{1}{2}$ $\frac{1}{2},\frac{1}{2},0$ $hkl: \; h+k,h+l,k+l=2n$

Symmetry of special projections

Along [001] $p2gm$ Along [111] $p6$ Along [110] $p2gg$

$\mathbf{a}'=\frac{1}{2}\mathbf{a}$ $\mathbf{b}'=\mathbf{b}$ $\mathbf{a}'=\frac{1}{3}(2\mathbf{a}-\mathbf{b}-\mathbf{c})$ $\mathbf{b}'=\frac{1}{3}(-\mathbf{a}+2\mathbf{b}-\mathbf{c})$ $\mathbf{a}'=\frac{1}{2}(-\mathbf{a}+\mathbf{b})$ $\mathbf{b}'=\mathbf{c}$

Origin at $0,0,z$ Origin at x,x,x Origin at $x,x,0$

Maximal non-isomorphic subgroups

I   [2] $P2_1 3$ (198)   1; 2; 3; 4; 5; 6; 7; 8; 9; 10; 11; 12

     [3] $Pa1$ ($Pbca$, 61)   1; 2; 3; 4; 13; 14; 15; 16

     ⎧ [4] $P1\bar{3}$ ($R\bar{3}$, 148)   1; 5; 9; 13; 17; 21

     ⎪ [4] $P1\bar{3}$ ($R\bar{3}$, 148)   1; 6; 12; 13; 18; 24

     ⎨ [4] $P1\bar{3}$ ($R\bar{3}$, 148)   1; 7; 10; 13; 19; 22

     ⎩ [4] $P1\bar{3}$ ($R\bar{3}$, 148)   1; 8; 11; 13; 20; 23

IIa   none

IIb   none

Maximal isomorphic subgroups of lowest index

IIc   [27] $Pa\bar{3}$ ($\mathbf{a}'=3\mathbf{a}, \mathbf{b}'=3\mathbf{b}, \mathbf{c}'=3\mathbf{c}$) (205)

Minimal non-isomorphic supergroups

I   none

II   [2] $Ia\bar{3}$ (206); [4] $Fm\bar{3}$ (202)

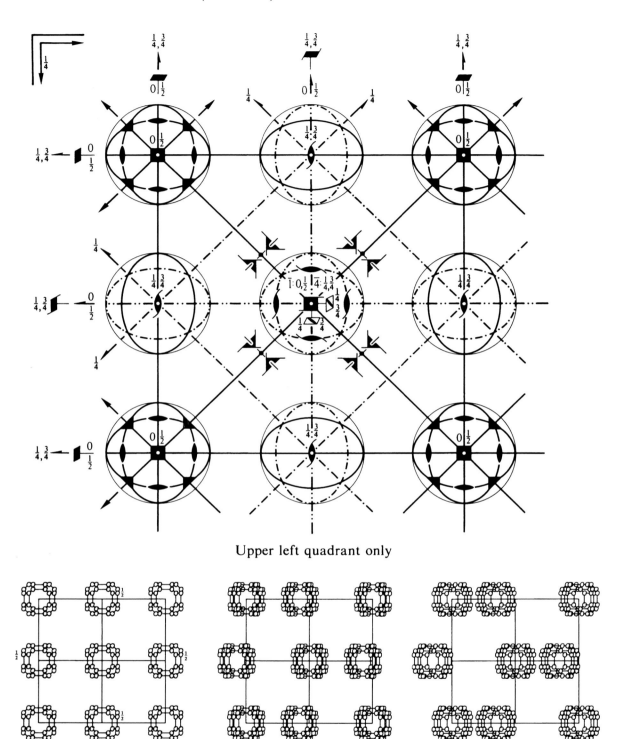

Upper left quadrant only

Origin at centre $(m\bar{3}m)$

Asymmetric unit $0 \le x \le \frac{1}{2}; \quad 0 \le y \le \frac{1}{4}; \quad 0 \le z \le \frac{1}{4}; \quad y \le \min(x, \frac{1}{2}-x); \quad z \le y$

       Vertices $0,0,0 \quad \frac{1}{2},0,0 \quad \frac{1}{4},\frac{1}{4},0 \quad \frac{1}{4},\frac{1}{4},\frac{1}{4}$

Symmetry operations

(given on page 153)

Generators selected (1); $t(1,0,0)$; $t(0,1,0)$; $t(0,0,1)$; $t(0,\frac{1}{2},\frac{1}{2})$; $t(\frac{1}{2},0,\frac{1}{2})$; (2); (3); (5); (13); (25)

Positions

Multiplicity, Wyckoff letter, Site symmetry	Coordinates $(0,0,0)+$ $(0,\frac{1}{2},\frac{1}{2})+$ $(\frac{1}{2},0,\frac{1}{2})+$ $(\frac{1}{2},\frac{1}{2},0)+$				Reflection conditions h,k,l permutable

192   i 1

(1) x,y,z (2) $\bar{x},\bar{y}+\frac{1}{2},z+\frac{1}{2}$ (3) $\bar{x}+\frac{1}{2},y+\frac{1}{2},\bar{z}$ (4) $x+\frac{1}{2},\bar{y},\bar{z}+\frac{1}{2}$

(5) z,x,y (6) $z+\frac{1}{2},\bar{x},\bar{y}+\frac{1}{2}$ (7) $\bar{z},\bar{x}+\frac{1}{2},y+\frac{1}{2}$ (8) $\bar{z}+\frac{1}{2},x+\frac{1}{2},\bar{y}$

(9) y,z,x (10) $\bar{y}+\frac{1}{2},z+\frac{1}{2},\bar{x}$ (11) $y+\frac{1}{2},\bar{z},\bar{x}+\frac{1}{2}$ (12) $\bar{y},\bar{z}+\frac{1}{2},x+\frac{1}{2}$

(13) $y+\frac{3}{4},x+\frac{1}{4},\bar{z}+\frac{3}{4}$ (14) $\bar{y}+\frac{1}{4},\bar{x}+\frac{1}{4},\bar{z}+\frac{1}{4}$ (15) $y+\frac{1}{4},\bar{x}+\frac{3}{4},z+\frac{3}{4}$ (16) $\bar{y}+\frac{3}{4},x+\frac{3}{4},z+\frac{1}{4}$

(17) $x+\frac{3}{4},z+\frac{1}{4},\bar{y}+\frac{3}{4}$ (18) $\bar{x}+\frac{3}{4},z+\frac{3}{4},y+\frac{1}{4}$ (19) $\bar{x}+\frac{1}{4},\bar{z}+\frac{1}{4},\bar{y}+\frac{1}{4}$ (20) $x+\frac{1}{4},\bar{z}+\frac{3}{4},y+\frac{3}{4}$

(21) $z+\frac{3}{4},y+\frac{1}{4},\bar{x}+\frac{3}{4}$ (22) $z+\frac{1}{4},\bar{y}+\frac{3}{4},x+\frac{3}{4}$ (23) $\bar{z}+\frac{3}{4},y+\frac{3}{4},x+\frac{1}{4}$ (24) $\bar{z}+\frac{1}{4},\bar{y}+\frac{1}{4},\bar{x}+\frac{1}{4}$

(25) $\bar{x}+\frac{1}{4},\bar{y}+\frac{1}{4},\bar{z}+\frac{1}{4}$ (26) $x+\frac{1}{4},y+\frac{3}{4},\bar{z}+\frac{3}{4}$ (27) $x+\frac{3}{4},\bar{y}+\frac{3}{4},z+\frac{1}{4}$ (28) $\bar{x}+\frac{3}{4},y+\frac{1}{4},z+\frac{3}{4}$

(29) $\bar{z}+\frac{1}{4},\bar{x}+\frac{1}{4},\bar{y}+\frac{1}{4}$ (30) $\bar{z}+\frac{3}{4},x+\frac{1}{4},y+\frac{3}{4}$ (31) $z+\frac{1}{4},x+\frac{3}{4},\bar{y}+\frac{3}{4}$ (32) $z+\frac{3}{4},\bar{x}+\frac{3}{4},y+\frac{1}{4}$

(33) $\bar{y}+\frac{1}{4},\bar{z}+\frac{1}{4},\bar{x}+\frac{1}{4}$ (34) $y+\frac{3}{4},\bar{z}+\frac{3}{4},x+\frac{1}{4}$ (35) $\bar{y}+\frac{3}{4},z+\frac{1}{4},x+\frac{3}{4}$ (36) $y+\frac{1}{4},z+\frac{3}{4},\bar{x}+\frac{3}{4}$

(37) $\bar{y}+\frac{1}{2},\bar{x},z+\frac{1}{2}$ (38) y,x,z (39) $\bar{y},x+\frac{1}{2},\bar{z}+\frac{1}{2}$ (40) $y+\frac{1}{2},\bar{x}+\frac{1}{2},\bar{z}$

(41) $\bar{x}+\frac{1}{2},\bar{z},y+\frac{1}{2}$ (42) $x+\frac{1}{2},\bar{z}+\frac{1}{2},\bar{y}$ (43) x,z,y (44) $\bar{x},z+\frac{1}{2},\bar{y}+\frac{1}{2}$

(45) $\bar{z}+\frac{1}{2},\bar{y},x+\frac{1}{2}$ (46) $\bar{z},y+\frac{1}{2},\bar{x}+\frac{1}{2}$ (47) $z+\frac{1}{2},\bar{y}+\frac{1}{2},\bar{x}$ (48) z,y,x

Reflection conditions for General:

hkl : $h+k=2n$ and $h+l,k+l=2n$
$0kl$: $k+l=4n$ and $k,l=2n$
hhl : $h+l=2n$
$h00$: $h=4n$

Special: as above, plus

96   h $..2$

$\frac{1}{8},y,\bar{y}+\frac{1}{4}$ $\frac{7}{8},\bar{y}+\frac{1}{2},\bar{y}+\frac{3}{4}$ $\frac{3}{8},y+\frac{1}{2},y+\frac{3}{4}$ $\frac{5}{8},\bar{y},y+\frac{1}{4}$
$\bar{y}+\frac{1}{4},\frac{1}{8},y$ $\bar{y}+\frac{3}{4},\frac{7}{8},\bar{y}+\frac{1}{2}$ $y+\frac{3}{4},\frac{3}{8},y+\frac{1}{2}$ $y+\frac{1}{4},\frac{5}{8},\bar{y}$
$y,\bar{y}+\frac{1}{4},\frac{1}{8}$ $\bar{y}+\frac{1}{2},\bar{y}+\frac{3}{4},\frac{7}{8}$ $y+\frac{1}{2},y+\frac{3}{4},\frac{3}{8}$ $\bar{y},y+\frac{1}{4},\frac{5}{8}$
$\frac{1}{8},\bar{y}+\frac{1}{4},y$ $\frac{3}{8},y+\frac{3}{4},y+\frac{1}{2}$ $\frac{7}{8},\bar{y}+\frac{3}{4},\bar{y}+\frac{1}{2}$ $\frac{5}{8},y+\frac{1}{4},\bar{y}$
$y,\frac{1}{8},\bar{y}+\frac{1}{4}$ $y+\frac{1}{2},\frac{3}{8},y+\frac{3}{4}$ $\bar{y}+\frac{1}{2},\frac{7}{8},\bar{y}+\frac{3}{4}$ $\bar{y},\frac{5}{8},y+\frac{1}{4}$
$\bar{y}+\frac{1}{4},y,\frac{1}{8}$ $y+\frac{3}{4},y+\frac{1}{2},\frac{3}{8}$ $\bar{y}+\frac{3}{4},\bar{y}+\frac{1}{2},\frac{7}{8}$ $y+\frac{1}{4},\bar{y},\frac{5}{8}$

no extra conditions

96   g $..m$

x,x,z $\bar{x},\bar{x}+\frac{1}{2},z+\frac{1}{2}$ $\bar{x}+\frac{1}{2},x+\frac{1}{2},\bar{z}$ $x+\frac{1}{2},\bar{x},\bar{z}+\frac{1}{2}$
z,x,x $z+\frac{1}{2},\bar{x},\bar{x}+\frac{1}{2}$ $\bar{z},\bar{x}+\frac{1}{2},x+\frac{1}{2}$ $\bar{z}+\frac{1}{2},x+\frac{1}{2},\bar{x}$
x,z,x $\bar{x}+\frac{1}{2},z+\frac{1}{2},\bar{x}$ $x+\frac{1}{2},\bar{z},\bar{x}+\frac{1}{2}$ $\bar{x},\bar{z}+\frac{1}{2},x+\frac{1}{2}$
$x+\frac{3}{4},x+\frac{1}{4},\bar{z}+\frac{3}{4}$ $\bar{x}+\frac{1}{4},\bar{x}+\frac{1}{4},\bar{z}+\frac{1}{4}$ $x+\frac{1}{4},\bar{x}+\frac{3}{4},z+\frac{3}{4}$ $\bar{x}+\frac{3}{4},x+\frac{3}{4},z+\frac{1}{4}$
$x+\frac{3}{4},z+\frac{1}{4},\bar{x}+\frac{3}{4}$ $\bar{x}+\frac{3}{4},z+\frac{3}{4},x+\frac{1}{4}$ $\bar{x}+\frac{1}{4},\bar{z}+\frac{1}{4},\bar{x}+\frac{1}{4}$ $x+\frac{1}{4},\bar{z}+\frac{3}{4},x+\frac{3}{4}$
$z+\frac{3}{4},x+\frac{1}{4},\bar{x}+\frac{3}{4}$ $z+\frac{1}{4},\bar{x}+\frac{3}{4},x+\frac{3}{4}$ $\bar{z}+\frac{3}{4},x+\frac{3}{4},x+\frac{1}{4}$ $\bar{z}+\frac{1}{4},\bar{x}+\frac{1}{4},\bar{x}+\frac{1}{4}$

no extra conditions

48   f $2.mm$

$x,0,0$ $\bar{x},\frac{1}{2},\frac{1}{2}$ $0,x,0$ $\frac{1}{2},\bar{x},\frac{1}{2}$ $0,0,x$ $\frac{1}{2},\frac{1}{2},\bar{x}$
$\frac{3}{4},x+\frac{1}{4},\frac{3}{4}$ $\frac{1}{4},\bar{x}+\frac{1}{4},\frac{1}{4}$ $x+\frac{3}{4},\frac{3}{4},\frac{1}{4}$ $\bar{x}+\frac{3}{4},\frac{3}{4},\frac{1}{4}$ $\frac{3}{4},\frac{1}{4},\bar{x}+\frac{3}{4}$ $\frac{1}{4},\frac{3}{4},x+\frac{3}{4}$

hkl : $h=2n+1$
or $h+k+l=4n$

32   e $.3m$

x,x,x $\bar{x},\bar{x}+\frac{1}{2},x+\frac{1}{2}$
$\bar{x}+\frac{1}{2},x+\frac{1}{2},\bar{x}$ $x+\frac{1}{2},\bar{x},\bar{x}+\frac{1}{2}$
$x+\frac{3}{4},x+\frac{1}{4},\bar{x}+\frac{3}{4}$ $\bar{x}+\frac{1}{4},\bar{x}+\frac{1}{4},\bar{x}+\frac{1}{4}$
$x+\frac{1}{4},\bar{x}+\frac{3}{4},x+\frac{3}{4}$ $\bar{x}+\frac{3}{4},x+\frac{3}{4},x+\frac{1}{4}$

no extra conditions

16   d $.\bar{3}m$

$\frac{5}{8},\frac{5}{8},\frac{5}{8}$ $\frac{3}{8},\frac{7}{8},\frac{1}{8}$ $\frac{7}{8},\frac{1}{8},\frac{3}{8}$ $\frac{1}{8},\frac{3}{8},\frac{7}{8}$

hkl : $h=2n+1$
or $h,k,l=4n+2$
or $h,k,l=4n$

16   c $.\bar{3}m$

$\frac{1}{8},\frac{1}{8},\frac{1}{8}$ $\frac{7}{8},\frac{3}{8},\frac{5}{8}$ $\frac{3}{8},\frac{5}{8},\frac{7}{8}$ $\frac{5}{8},\frac{7}{8},\frac{3}{8}$

8   b $\bar{4}3m$

$\frac{1}{2},\frac{1}{2},\frac{1}{2}$ $\frac{1}{4},\frac{3}{4},\frac{1}{4}$

hkl : $h=2n+1$
or $h+k+l=4n$

8   a $\bar{4}3m$

$0,0,0$ $\frac{3}{4},\frac{1}{4},\frac{3}{4}$

Symmetry of special projections

Along [001] $p4mm$
$\mathbf{a}' = \frac{1}{4}(\mathbf{a}-\mathbf{b})$ $\mathbf{b}' = \frac{1}{4}(\mathbf{a}+\mathbf{b})$
Origin at $0,0,z$

Along [111] $p6mm$
$\mathbf{a}' = \frac{1}{6}(2\mathbf{a}-\mathbf{b}-\mathbf{c})$ $\mathbf{b}' = \frac{1}{6}(-\mathbf{a}+2\mathbf{b}-\mathbf{c})$
Origin at x,x,x

Along [110] $c2mm$
$\mathbf{a}' = \frac{1}{2}(-\mathbf{a}+\mathbf{b})$ $\mathbf{b}' = \mathbf{c}$
Origin at $x,x,\frac{1}{8}$

ORIGIN CHOICE 1

Maximal non-isomorphic subgroups

I   [2] $F \bar{4} 3 m$ (216)   (1; 2; 3; 4; 5; 6; 7; 8; 9; 10; 11; 12; 37; 38; 39; 40; 41; 42; 43; 44; 45; 46; 47; 48)+

       [2] $F 4_1 3 2$ (210)   (1; 2; 3; 4; 5; 6; 7; 8; 9; 10; 11; 12; 13; 14; 15; 16; 17; 18; 19; 20; 21; 22; 23; 24)+

       [2] $F d \bar{3} 1$ ($F d \bar{3}$, 203)   (1; 2; 3; 4; 5; 6; 7; 8; 9; 10; 11; 12; 25; 26; 27; 28; 29; 30; 31; 32; 33; 34; 35; 36)+

       $\begin{cases} \text{[3] } F 4_1/d 1 2/m \text{ (} I 4_1/a m d, 141) \\ \text{[3] } F 4_1/d 1 2/m \text{ (} I 4_1/a m d, 141) \\ \text{[3] } F 4_1/d 1 2/m \text{ (} I 4_1/a m d, 141) \end{cases}$ (1; 2; 3; 4; 13; 14; 15; 16; 25; 26; 27; 28; 37; 38; 39; 40)+

       (1; 2; 3; 4; 17; 18; 19; 20; 25; 26; 27; 28; 41; 42; 43; 44)+

       (1; 2; 3; 4; 21; 22; 23; 24; 25; 26; 27; 28; 45; 46; 47; 48)+

       $\begin{cases} \text{[4] } F 1 \bar{3} 2/m \text{ (} R \bar{3} m, 166) \\ \text{[4] } F 1 \bar{3} 2/m \text{ (} R \bar{3} m, 166) \\ \text{[4] } F 1 \bar{3} 2/m \text{ (} R \bar{3} m, 166) \\ \text{[4] } F 1 \bar{3} 2/m \text{ (} R \bar{3} m, 166) \end{cases}$ (1; 5; 9; 14; 19; 24; 25; 29; 33; 38; 43; 48)+

       (1; 6; 12; 13; 18; 24; 25; 30; 36; 37; 42; 48)+

       (1; 7; 10; 13; 19; 22; 25; 31; 34; 37; 43; 46)+

       (1; 8; 11; 14; 18; 22; 25; 32; 35; 38; 42; 46)+

IIa   none

IIb   none

Maximal isomorphic subgroups of lowest index

IIc   [27] $F d \bar{3} m$ ($\mathbf{a}' = 3\mathbf{a}, \mathbf{b}' = 3\mathbf{b}, \mathbf{c}' = 3\mathbf{c}$) (227)

Minimal non-isomorphic supergroups

I   none

II   [2] $P n \bar{3} m$ ($\mathbf{a}' = \frac{1}{2}\mathbf{a}, \mathbf{b}' = \frac{1}{2}\mathbf{b}, \mathbf{c}' = \frac{1}{2}\mathbf{c}$) (224)

(*from page* 154)

Symmetry operations

For $(0,0,0)+$ set

(1) 1
(2) $2(0,0,\frac{1}{2})\quad 0,\frac{1}{4},z$
(3) $2(0,\frac{1}{2},0)\quad \frac{1}{4},y,0$
(4) $2(\frac{1}{2},0,0)\quad x,0,\frac{1}{4}$
(5) $3^{+}\quad x,x,x$
(6) $3^{+}(\frac{1}{3},-\frac{1}{3},\frac{1}{3})\quad \bar{x}+\frac{1}{6},x+\frac{1}{6},\bar{x}$
(7) $3^{+}(-\frac{1}{3},\frac{1}{3},\frac{1}{3})\quad x+\frac{1}{3},\bar{x}-\frac{1}{6},\bar{x}$
(8) $3^{+}(\frac{1}{3},\frac{1}{3},-\frac{1}{3})\quad \bar{x}+\frac{1}{6},\bar{x}+\frac{1}{3},x$
(9) $3^{-}\quad x,x,x$
(10) $3^{-}\quad x,\bar{x}+\frac{1}{2},\bar{x}$
(11) $3^{-}\quad \bar{x}+\frac{1}{2},\bar{x},x$
(12) $3^{-}\quad \bar{x}-\frac{1}{2},x+\frac{1}{2},\bar{x}$
(13) $2(\frac{1}{2},\frac{1}{2},0)\quad x,x-\frac{1}{4},\frac{3}{8}$
(14) $2\quad x,\bar{x}+\frac{1}{4},\frac{1}{8}$
(15) $4^{-}(0,0,\frac{3}{4})\quad \frac{1}{2},\frac{1}{4},z$
(16) $4^{+}(0,0,\frac{1}{4})\quad 0,\frac{3}{4},z$
(17) $4^{-}(\frac{3}{4},0,0)\quad x,\frac{1}{2},\frac{1}{4}$
(18) $2(0,\frac{1}{2},\frac{1}{2})\quad \frac{3}{8},y+\frac{1}{4},y$
(19) $2\quad \frac{1}{8},y+\frac{1}{4},\bar{y}$
(20) $4^{+}(\frac{1}{4},0,0)\quad x,0,\frac{3}{4}$
(21) $4^{+}(0,\frac{1}{4},0)\quad \frac{3}{4},y,0$
(22) $2(\frac{1}{2},0,\frac{1}{2})\quad x-\frac{1}{4},\frac{3}{8},x$
(23) $4^{-}(0,\frac{3}{4},0)\quad \frac{1}{4},y,\frac{1}{2}$
(24) $2\quad \bar{x}+\frac{1}{4},\frac{1}{8},x$
(25) $\bar{1}\quad \frac{1}{8},\frac{1}{8},\frac{1}{8}$
(26) $d(\frac{1}{4},\frac{3}{4},0)\quad x,y,\frac{3}{8}$
(27) $d(\frac{3}{4},0,\frac{1}{4})\quad x,\frac{3}{8},z$
(28) $d(0,\frac{1}{4},\frac{3}{4})\quad \frac{3}{8},y,z$
(29) $\bar{3}^{+}\quad x,x,x;\ \frac{1}{8},\frac{1}{8},\frac{1}{8}$
(30) $\bar{3}^{+}\quad \bar{x}-1,x+1,\bar{x};\ -\frac{1}{8},\frac{1}{8},\frac{7}{8}$
(31) $\bar{3}^{+}\quad x,\bar{x}+1,\bar{x};\ \frac{1}{8},\frac{7}{8},-\frac{1}{8}$
(32) $\bar{3}^{+}\quad \bar{x}+1,\bar{x},x;\ \frac{7}{8},-\frac{1}{8},\frac{1}{8}$
(33) $\bar{3}^{-}\quad x,x,x;\ \frac{1}{8},\frac{1}{8},\frac{1}{8}$
(34) $\bar{3}^{-}\quad x+\frac{3}{2},\bar{x}-1,\bar{x};\ \frac{5}{8},-\frac{1}{8},\frac{7}{8}$
(35) $\bar{3}^{-}\quad \bar{x}+\frac{1}{2},\bar{x}+\frac{3}{2},x;\ -\frac{1}{8},\frac{7}{8},\frac{5}{8}$
(36) $\bar{3}^{-}\quad \bar{x}+1,x+\frac{1}{2},\bar{x};\ \frac{7}{8},\frac{5}{8},-\frac{1}{8}$
(37) $g(\frac{1}{4},-\frac{1}{4},\frac{1}{2})\quad x+\frac{1}{4},\bar{x},z$
(38) $m\quad x,x,z$
(39) $\bar{4}^{-}\quad -\frac{1}{4},\frac{1}{4},z;\ -\frac{1}{4},\frac{1}{4},\frac{1}{4}$
(40) $\bar{4}^{+}\quad \frac{1}{2},0,z;\ \frac{1}{2},0,0$
(41) $\bar{4}^{-}\quad x,-\frac{1}{4},\frac{1}{4};\ \frac{1}{4},-\frac{1}{4},\frac{1}{4}$
(42) $g(\frac{1}{2},\frac{1}{4},-\frac{1}{4})\quad x,y+\frac{1}{4},\bar{y}$
(43) $m\quad x,y,y$
(44) $\bar{4}^{+}\quad x,\frac{1}{2},0;\ 0,\frac{1}{2},0$
(45) $\bar{4}^{+}\quad 0,y,\frac{1}{2};\ 0,0,\frac{1}{2}$
(46) $g(-\frac{1}{4},\frac{1}{2},\frac{1}{4})\quad \bar{x}+\frac{1}{4},y,x$
(47) $\bar{4}^{-}\quad \frac{1}{4},y,-\frac{1}{4};\ \frac{1}{4},\frac{1}{4},-\frac{1}{4}$
(48) $m\quad x,y,x$

For $(0,\frac{1}{2},\frac{1}{2})+$ set

(1) $t(0,\frac{1}{2},\frac{1}{2})$
(2) $2\quad 0,0,z$
(3) $2\quad \frac{1}{4},y,\frac{1}{4}$
(4) $2(\frac{1}{2},0,0)\quad x,\frac{1}{4},0$
(5) $3^{+}(\frac{1}{3},\frac{1}{3},\frac{1}{3})\quad x-\frac{1}{3},x-\frac{1}{6},x$
(6) $3^{+}\quad \bar{x}+\frac{1}{2},x,\bar{x}$
(7) $3^{+}\quad x,\bar{x},\bar{x}$
(8) $3^{+}\quad \bar{x}+\frac{1}{2},\bar{x}+\frac{1}{2},x$
(9) $3^{-}(\frac{1}{3},\frac{1}{3},\frac{1}{3})\quad x-\frac{1}{6},x+\frac{1}{6},x$
(10) $3^{-}\quad x+\frac{1}{2},\bar{x},\bar{x}$
(11) $3^{-}(\frac{1}{3},\frac{1}{3},-\frac{1}{3})\quad \bar{x}+\frac{1}{3},\bar{x}+\frac{1}{6},x$
(12) $3^{-}\quad \bar{x},x,\bar{x}$
(13) $2(\frac{3}{4},\frac{3}{4},0)\quad x,x,\frac{1}{8}$
(14) $2(-\frac{1}{4},\frac{1}{4},0)\quad x,\bar{x}+\frac{1}{2},\frac{3}{8}$
(15) $4^{-}(0,0,\frac{1}{4})\quad \frac{1}{4},0,z$
(16) $4^{+}(0,0,\frac{3}{4})\quad \frac{1}{4},\frac{1}{2},z$
(17) $4^{-}(\frac{3}{4},0,0)\quad x,\frac{1}{2},-\frac{1}{4}$
(18) $2(0,\frac{1}{2},\frac{1}{2})\quad \frac{3}{8},y-\frac{1}{4},y$
(19) $2\quad \frac{1}{8},y+\frac{3}{4},\bar{y}$
(20) $4^{+}(\frac{1}{4},0,0)\quad x,0,\frac{1}{4}$
(21) $4^{+}(0,\frac{3}{4},0)\quad \frac{1}{2},y,-\frac{1}{4}$
(22) $2(\frac{1}{4},0,\frac{1}{4})\quad x,\frac{1}{8},x$
(23) $4^{-}(0,\frac{1}{4},0)\quad 0,y,\frac{3}{4}$
(24) $2(-\frac{1}{4},0,\frac{1}{4})\quad \bar{x}+\frac{1}{2},\frac{3}{8},x$
(25) $\bar{1}\quad \frac{1}{8},\frac{3}{8},\frac{3}{8}$
(26) $d(\frac{1}{4},\frac{1}{4},0)\quad x,y,\frac{1}{8}$
(27) $d(\frac{3}{4},0,\frac{3}{4})\quad x,\frac{1}{8},z$
(28) $d(0,\frac{3}{4},\frac{1}{4})\quad \frac{3}{8},y,z$
(29) $\bar{3}^{+}\quad x,x+\frac{1}{2},x;\ \frac{1}{8},\frac{5}{8},\frac{1}{8}$
(30) $\bar{3}^{+}\quad \bar{x}-1,x+\frac{3}{2},\bar{x};\ -\frac{1}{8},\frac{5}{8},\frac{7}{8}$
(31) $\bar{3}^{+}\quad x,\bar{x}+\frac{1}{2},\bar{x};\ \frac{1}{8},\frac{3}{8},-\frac{1}{8}$
(32) $\bar{3}^{+}\quad \bar{x}+1,\bar{x}-\frac{1}{2},x;\ \frac{7}{8},-\frac{5}{8},\frac{1}{8}$
(33) $\bar{3}^{-}\quad x-\frac{1}{2},x-\frac{1}{2},x;\ \frac{1}{8},\frac{1}{8},\frac{5}{8}$
(34) $\bar{3}^{-}\quad x+1,\bar{x}-\frac{3}{2},\bar{x};\ \frac{1}{8},-\frac{5}{8},\frac{7}{8}$
(35) $\bar{3}^{-}\quad \bar{x},\bar{x}+1,x;\ -\frac{1}{8},\frac{7}{8},\frac{1}{8}$
(36) $\bar{3}^{-}\quad \bar{x}+\frac{1}{2},x,\bar{x};\ \frac{3}{8},\frac{1}{8},-\frac{1}{8}$
(37) $m\quad x+\frac{1}{2},\bar{x},z$
(38) $g(\frac{1}{4},\frac{1}{4},\frac{1}{2})\quad x-\frac{1}{4},x,z$
(39) $\bar{4}^{-}\quad 0,0,z;\ 0,0,0$
(40) $\bar{4}^{+}\quad \frac{1}{4},-\frac{1}{4},z;\ \frac{1}{4},-\frac{1}{4},\frac{1}{4}$
(41) $\bar{4}^{-}\quad x,\frac{1}{4},\frac{1}{4};\ \frac{1}{4},\frac{1}{4},\frac{1}{4}$
(42) $g(\frac{1}{2},-\frac{1}{4},\frac{1}{4})\quad x,y+\frac{1}{4},\bar{y}$
(43) $g(0,\frac{1}{2},\frac{1}{2})\quad x,y,y$
(44) $\bar{4}^{+}\quad x,0,0;\ 0,0,0$
(45) $\bar{4}^{+}\quad \frac{1}{4},y,\frac{1}{4};\ \frac{1}{4},\frac{1}{4},\frac{1}{4}$
(46) $m\quad \bar{x},y,x$
(47) $\bar{4}^{-}\quad \frac{1}{2},y,0;\ \frac{1}{2},0,0$
(48) $g(\frac{1}{4},\frac{1}{2},\frac{1}{4})\quad x-\frac{1}{4},y,x$

For $(\frac{1}{2},0,\frac{1}{2})+$ set

(1) $t(\frac{1}{2},0,\frac{1}{2})$
(2) $2\quad \frac{1}{4},\frac{1}{4},z$
(3) $2(0,\frac{1}{2},0)\quad 0,y,\frac{1}{4}$
(4) $2\quad x,0,0$
(5) $3^{+}(\frac{1}{3},\frac{1}{3},\frac{1}{3})\quad x+\frac{1}{6},x-\frac{1}{6},x$
(6) $3^{+}\quad \bar{x},x,\bar{x}$
(7) $3^{+}\quad x+\frac{1}{2},\bar{x},\bar{x}$
(8) $3^{+}\quad \bar{x},\bar{x}+\frac{1}{2},x$
(9) $3^{-}(\frac{1}{3},\frac{1}{3},\frac{1}{3})\quad x-\frac{1}{6},x-\frac{1}{3},x$
(10) $3^{-}(-\frac{1}{3},\frac{1}{3},\frac{1}{3})\quad x+\frac{1}{6},\bar{x}+\frac{1}{6},\bar{x}$
(11) $3^{-}\quad \bar{x},\bar{x},x$
(12) $3^{-}\quad \bar{x},x+\frac{1}{2},\bar{x}$
(13) $2(\frac{1}{4},\frac{1}{4},0)\quad x,x,\frac{1}{8}$
(14) $2(\frac{1}{4},-\frac{1}{4},0)\quad x,\bar{x}+\frac{1}{2},\frac{3}{8}$
(15) $4^{-}(0,0,\frac{1}{4})\quad \frac{3}{4},0,z$
(16) $4^{+}(0,0,\frac{3}{4})\quad -\frac{1}{4},\frac{1}{2},z$
(17) $4^{-}(\frac{1}{4},0,0)\quad x,\frac{1}{4},0$
(18) $2(0,\frac{3}{4},\frac{3}{4})\quad \frac{1}{8},y,y$
(19) $2(0,-\frac{1}{4},\frac{1}{4})\quad \frac{3}{8},y+\frac{1}{2},\bar{y}$
(20) $4^{+}(\frac{3}{4},0,0)\quad x,\frac{1}{4},\frac{1}{2}$
(21) $4^{+}(0,\frac{1}{4},0)\quad \frac{1}{4},y,0$
(22) $2(\frac{1}{2},0,\frac{1}{2})\quad x+\frac{1}{4},\frac{3}{8},x$
(23) $4^{-}(0,\frac{3}{4},0)\quad -\frac{1}{4},y,\frac{1}{2}$
(24) $2\quad \bar{x}+\frac{3}{4},\frac{1}{8},x$
(25) $\bar{1}\quad \frac{3}{8},\frac{1}{8},\frac{3}{8}$
(26) $d(\frac{3}{4},\frac{3}{4},0)\quad x,y,\frac{1}{8}$
(27) $d(\frac{1}{4},0,\frac{3}{4})\quad x,\frac{3}{8},z$
(28) $d(0,\frac{1}{4},\frac{1}{4})\quad \frac{1}{8},y,z$
(29) $\bar{3}^{+}\quad x-\frac{1}{2},x-\frac{1}{2},x;\ \frac{1}{8},\frac{1}{8},\frac{5}{8}$
(30) $\bar{3}^{+}\quad \bar{x}-\frac{1}{2},x+\frac{1}{2},\bar{x};\ -\frac{1}{8},\frac{1}{8},\frac{3}{8}$
(31) $\bar{3}^{+}\quad x-\frac{1}{2},\bar{x}+\frac{3}{2},\bar{x};\ \frac{1}{8},\frac{7}{8},-\frac{5}{8}$
(32) $\bar{3}^{+}\quad \bar{x}+\frac{3}{2},\bar{x}+\frac{1}{2},x;\ \frac{7}{8},-\frac{1}{8},\frac{5}{8}$
(33) $\bar{3}^{-}\quad x+\frac{1}{2},x,x;\ \frac{5}{8},\frac{1}{8},\frac{1}{8}$
(34) $\bar{3}^{-}\quad x+1,\bar{x}-1,\bar{x};\ \frac{1}{8},-\frac{1}{8},\frac{7}{8}$
(35) $\bar{3}^{-}\quad \bar{x},\bar{x}+\frac{1}{2},x;\ -\frac{1}{8},\frac{3}{8},\frac{1}{8}$
(36) $\bar{3}^{-}\quad \bar{x}+\frac{3}{2},x-\frac{1}{2},\bar{x};\ \frac{7}{8},\frac{1}{8},-\frac{5}{8}$
(37) $m\quad x,\bar{x},z$
(38) $g(\frac{1}{4},\frac{1}{4},\frac{1}{2})\quad x+\frac{1}{4},x,z$
(39) $\bar{4}^{-}\quad 0,\frac{1}{2},z;\ 0,\frac{1}{2},0$
(40) $\bar{4}^{+}\quad \frac{1}{4},\frac{1}{4},z;\ \frac{1}{4},\frac{1}{4},\frac{1}{4}$
(41) $\bar{4}^{-}\quad x,0,0;\ 0,0,0$
(42) $m\quad x,y+\frac{1}{2},\bar{y}$
(43) $g(\frac{1}{2},\frac{1}{4},\frac{1}{4})\quad x,y-\frac{1}{4},y$
(44) $\bar{4}^{+}\quad x,\frac{1}{4},-\frac{1}{4};\ \frac{1}{4},\frac{1}{4},-\frac{1}{4}$
(45) $\bar{4}^{+}\quad 0,y,0;\ 0,0,0$
(46) $g(\frac{1}{4},\frac{1}{2},-\frac{1}{4})\quad \bar{x}+\frac{1}{4},y,x$
(47) $\bar{4}^{-}\quad \frac{1}{4},y,\frac{1}{4};\ \frac{1}{4},\frac{1}{4},\frac{1}{4}$
(48) $g(\frac{1}{2},0,\frac{1}{2})\quad x,y,x$

For $(\frac{1}{2},\frac{1}{2},0)+$ set

(1) $t(\frac{1}{2},\frac{1}{2},0)$
(2) $2(0,0,\frac{1}{2})\quad \frac{1}{4},0,z$
(3) $2\quad 0,y,0$
(4) $2\quad x,\frac{1}{4},\frac{1}{4}$
(5) $3^{+}(\frac{1}{3},\frac{1}{3},\frac{1}{3})\quad x+\frac{1}{6},x+\frac{1}{3},x$
(6) $3^{+}\quad \bar{x},x+\frac{1}{2},\bar{x}$
(7) $3^{+}\quad x+\frac{1}{2},\bar{x}-\frac{1}{2},\bar{x}$
(8) $3^{+}\quad \bar{x},\bar{x},x$
(9) $3^{-}(\frac{1}{3},\frac{1}{3},\frac{1}{3})\quad x+\frac{1}{3},x+\frac{1}{6},x$
(10) $3^{-}\quad x,\bar{x},\bar{x}$
(11) $3^{-}\quad \bar{x}+\frac{1}{2},\bar{x}+\frac{1}{2},x$
(12) $3^{-}(\frac{1}{3},-\frac{1}{3},\frac{1}{3})\quad \bar{x}-\frac{1}{6},x+\frac{1}{3},\bar{x}$
(13) $2(\frac{1}{2},\frac{1}{2},0)\quad x,x+\frac{1}{4},\frac{3}{8}$
(14) $2\quad x,\bar{x}+\frac{3}{4},\frac{1}{8}$
(15) $4^{-}(0,0,\frac{3}{4})\quad \frac{1}{2},-\frac{1}{4},z$
(16) $4^{+}(0,0,\frac{1}{4})\quad 0,\frac{1}{4},z$
(17) $4^{-}(\frac{1}{4},0,0)\quad x,\frac{3}{4},0$
(18) $2(0,\frac{1}{4},\frac{1}{4})\quad \frac{1}{8},y,y$
(19) $2(0,\frac{1}{4},-\frac{1}{4})\quad \frac{3}{8},y+\frac{1}{2},\bar{y}$
(20) $4^{+}(\frac{3}{4},0,0)\quad x,-\frac{1}{4},\frac{1}{2}$
(21) $4^{+}(0,\frac{3}{4},0)\quad \frac{1}{2},y,\frac{1}{4}$
(22) $2(\frac{3}{4},0,\frac{3}{4})\quad x,\frac{1}{8},x$
(23) $4^{-}(0,\frac{1}{4},0)\quad 0,y,\frac{1}{4}$
(24) $2(\frac{1}{4},0,-\frac{1}{4})\quad \bar{x}+\frac{1}{2},\frac{3}{8},x$
(25) $\bar{1}\quad \frac{3}{8},\frac{3}{8},\frac{1}{8}$
(26) $d(\frac{3}{4},\frac{1}{4},0)\quad x,y,\frac{3}{8}$
(27) $d(\frac{1}{4},0,\frac{1}{4})\quad x,\frac{1}{8},z$
(28) $d(0,\frac{3}{4},\frac{3}{4})\quad \frac{1}{8},y,z$
(29) $\bar{3}^{+}\quad x+\frac{1}{2},x,x;\ \frac{5}{8},\frac{1}{8},\frac{1}{8}$
(30) $\bar{3}^{+}\quad \bar{x}-\frac{3}{2},x+1,\bar{x};\ -\frac{5}{8},\frac{1}{8},\frac{7}{8}$
(31) $\bar{3}^{+}\quad x+\frac{1}{2},\bar{x}+1,\bar{x};\ \frac{5}{8},\frac{7}{8},-\frac{1}{8}$
(32) $\bar{3}^{+}\quad \bar{x}+\frac{1}{2},\bar{x},x;\ \frac{3}{8},-\frac{1}{8},\frac{1}{8}$
(33) $\bar{3}^{-}\quad x,x+\frac{1}{2},x;\ \frac{1}{8},\frac{5}{8},\frac{1}{8}$
(34) $\bar{3}^{-}\quad x+\frac{1}{2},\bar{x}-\frac{1}{2},\bar{x};\ \frac{1}{8},-\frac{1}{8},\frac{3}{8}$
(35) $\bar{3}^{-}\quad \bar{x}-\frac{1}{2},\bar{x}+1,x;\ -\frac{5}{8},\frac{7}{8},\frac{1}{8}$
(36) $\bar{3}^{-}\quad \bar{x}+1,x,\bar{x};\ \frac{7}{8},\frac{1}{8},-\frac{1}{8}$
(37) $g(-\frac{1}{4},\frac{1}{4},\frac{1}{2})\quad x+\frac{1}{4},\bar{x},z$
(38) $g(\frac{1}{2},\frac{1}{2},0)\quad x,x,z$
(39) $\bar{4}^{-}\quad \frac{1}{4},\frac{1}{4},z;\ \frac{1}{4},\frac{1}{4},\frac{1}{4}$
(40) $\bar{4}^{+}\quad 0,0,z;\ 0,0,0$
(41) $\bar{4}^{-}\quad x,0,\frac{1}{2};\ 0,0,\frac{1}{2}$
(42) $m\quad x,y,\bar{y}$
(43) $g(\frac{1}{2},\frac{1}{4},\frac{1}{4})\quad x,y+\frac{1}{4},y$
(44) $\bar{4}^{+}\quad x,\frac{1}{4},\frac{1}{4};\ \frac{1}{4},\frac{1}{4},\frac{1}{4}$
(45) $\bar{4}^{+}\quad -\frac{1}{4},y,\frac{1}{4};\ -\frac{1}{4},\frac{1}{4},\frac{1}{4}$
(46) $m\quad \bar{x}+\frac{1}{2},y,x$
(47) $\bar{4}^{-}\quad 0,y,0;\ 0,0,0$
(48) $g(\frac{1}{4},\frac{1}{2},\frac{1}{4})\quad x+\frac{1}{4},y,x$

$F\,d\,\bar{3}\,m$

O_h^7

$m\,\bar{3}\,m$

Cubic

No. 227

$F\;4_1/d\;\bar{3}\;2/m$

Patterson symmetry $F\,m\,\bar{3}\,m$

ORIGIN CHOICE 2

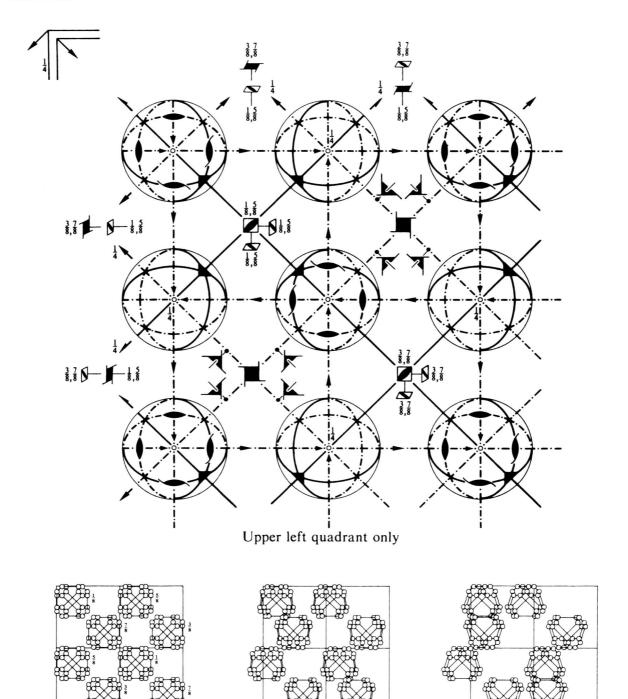

Upper left quadrant only

Origin at centre $(\bar{3}m)$, at $\tfrac{1}{8},\tfrac{1}{8},\tfrac{1}{8}$ from $\bar{4}3m$

Asymmetric unit $\quad -\tfrac{1}{8} \le x \le \tfrac{3}{8}; \quad -\tfrac{1}{8} \le y \le 0; \quad -\tfrac{1}{4} \le z \le 0; \quad y \le \min(\tfrac{1}{4} - x, x); \quad -y - \tfrac{1}{4} \le z \le y$

Vertices $\quad -\tfrac{1}{8}, -\tfrac{1}{8}, -\tfrac{1}{8} \quad \tfrac{3}{8}, -\tfrac{1}{8}, -\tfrac{1}{8} \quad \tfrac{1}{4}, 0, 0 \quad 0, 0, 0 \quad \tfrac{1}{4}, 0, -\tfrac{1}{4} \quad 0, 0, -\tfrac{1}{4}$

Symmetry operations

(*given on page* 161)

Generators selected (1); $t(1,0,0)$; $t(0,1,0)$; $t(0,0,1)$; $t(0,\frac{1}{2},\frac{1}{2})$; $t(\frac{1}{2},0,\frac{1}{2})$; (2); (3); (5); (13); (25)

Positions

Multiplicity, Wyckoff letter, Site symmetry	Coordinates $(0,0,0)+$ $(0,\frac{1}{2},\frac{1}{2})+$ $(\frac{1}{2},0,\frac{1}{2})+$ $(\frac{1}{2},\frac{1}{2},0)+$	Reflection conditions h,k,l permutable

General:

192   i 1

(1) x,y,z (2) $\bar{x}+\frac{3}{4},\bar{y}+\frac{1}{4},z+\frac{1}{2}$ (3) $\bar{x}+\frac{1}{4},y+\frac{1}{2},\bar{z}+\frac{3}{4}$ (4) $x+\frac{1}{2},\bar{y}+\frac{3}{4},\bar{z}+\frac{1}{4}$

(5) z,x,y (6) $z+\frac{1}{2},\bar{x}+\frac{3}{4},\bar{y}+\frac{1}{4}$ (7) $\bar{z}+\frac{3}{4},\bar{x}+\frac{1}{4},y+\frac{1}{2}$ (8) $\bar{z}+\frac{1}{4},x+\frac{1}{2},\bar{y}+\frac{3}{4}$

(9) y,z,x (10) $\bar{y}+\frac{1}{4},z+\frac{1}{2},\bar{x}+\frac{3}{4}$ (11) $y+\frac{1}{2},\bar{z}+\frac{3}{4},\bar{x}+\frac{1}{4}$ (12) $\bar{y}+\frac{3}{4},\bar{z}+\frac{1}{4},x+\frac{1}{2}$

(13) $y+\frac{3}{4},x+\frac{1}{4},\bar{z}+\frac{1}{2}$ (14) $\bar{y},\bar{x},\bar{z}$ (15) $y+\frac{1}{4},\bar{x}+\frac{1}{2},z+\frac{3}{4}$ (16) $\bar{y}+\frac{1}{2},x+\frac{3}{4},z+\frac{1}{4}$

(17) $x+\frac{3}{4},z+\frac{1}{4},\bar{y}+\frac{1}{2}$ (18) $\bar{x}+\frac{1}{2},z+\frac{3}{4},y+\frac{1}{4}$ (19) $\bar{x},\bar{z},\bar{y}$ (20) $x+\frac{1}{4},\bar{z}+\frac{1}{2},y+\frac{3}{4}$

(21) $z+\frac{3}{4},y+\frac{1}{4},\bar{x}+\frac{1}{2}$ (22) $z+\frac{1}{4},\bar{y}+\frac{1}{2},x+\frac{3}{4}$ (23) $\bar{z}+\frac{1}{2},y+\frac{3}{4},x+\frac{1}{4}$ (24) $\bar{z},\bar{y},\bar{x}$

(25) $\bar{x},\bar{y},\bar{z}$ (26) $x+\frac{1}{4},y+\frac{3}{4},\bar{z}+\frac{1}{2}$ (27) $x+\frac{3}{4},\bar{y}+\frac{1}{2},z+\frac{1}{4}$ (28) $\bar{x}+\frac{1}{2},y+\frac{1}{4},z+\frac{3}{4}$

(29) $\bar{z},\bar{x},\bar{y}$ (30) $\bar{z}+\frac{1}{2},x+\frac{1}{4},y+\frac{3}{4}$ (31) $z+\frac{1}{4},x+\frac{3}{4},\bar{y}+\frac{1}{2}$ (32) $z+\frac{3}{4},\bar{x}+\frac{1}{2},y+\frac{1}{4}$

(33) $\bar{y},\bar{z},\bar{x}$ (34) $y+\frac{3}{4},\bar{z}+\frac{1}{2},x+\frac{1}{4}$ (35) $\bar{y}+\frac{1}{2},z+\frac{1}{4},x+\frac{3}{4}$ (36) $y+\frac{1}{4},z+\frac{3}{4},\bar{x}+\frac{1}{2}$

(37) $\bar{y}+\frac{1}{4},\bar{x}+\frac{3}{4},z+\frac{1}{2}$ (38) y,x,z (39) $\bar{y}+\frac{3}{4},x+\frac{1}{2},\bar{z}+\frac{1}{4}$ (40) $y+\frac{1}{2},\bar{x}+\frac{1}{4},\bar{z}+\frac{3}{4}$

(41) $\bar{x}+\frac{1}{4},\bar{z}+\frac{3}{4},y+\frac{1}{2}$ (42) $x+\frac{1}{2},\bar{z}+\frac{1}{4},\bar{y}+\frac{3}{4}$ (43) x,z,y (44) $\bar{x}+\frac{3}{4},z+\frac{1}{2},\bar{y}+\frac{1}{4}$

(45) $\bar{z}+\frac{1}{4},\bar{y}+\frac{3}{4},x+\frac{1}{2}$ (46) $\bar{z}+\frac{3}{4},y+\frac{1}{2},\bar{x}+\frac{1}{4}$ (47) $z+\frac{1}{2},\bar{y}+\frac{1}{4},\bar{x}+\frac{3}{4}$ (48) z,y,x

Reflection conditions:

hkl : $h+k=2n$ and $h+l,k+l=2n$
$0kl$: $k+l=4n$ and $k,l=2n$
hhl : $h+l=2n$
$h00$: $h=4n$

Special: as above, plus

96   h $..2$

$0,y,\bar{y}$ $\frac{3}{4},\bar{y}+\frac{1}{4},\bar{y}+\frac{1}{2}$ $\frac{1}{4},y+\frac{1}{2},y+\frac{3}{4}$ $\frac{1}{2},\bar{y}+\frac{3}{4},y+\frac{1}{4}$

$\bar{y},0,y$ $\bar{y}+\frac{1}{2},\frac{3}{4},\bar{y}+\frac{1}{4}$ $y+\frac{3}{4},\frac{1}{4},y+\frac{1}{2}$ $y+\frac{1}{4},\frac{1}{2},\bar{y}+\frac{3}{4}$

$y,\bar{y},0$ $\bar{y}+\frac{1}{4},\bar{y}+\frac{1}{2},\frac{3}{4}$ $y+\frac{1}{2},y+\frac{3}{4},\frac{1}{4}$ $\bar{y}+\frac{3}{4},y+\frac{1}{4},\frac{1}{2}$

$0,\bar{y},y$ $\frac{1}{4},y+\frac{3}{4},y+\frac{1}{2}$ $\frac{3}{4},\bar{y}+\frac{1}{2},\bar{y}+\frac{1}{4}$ $\frac{1}{2},y+\frac{1}{4},\bar{y}+\frac{3}{4}$

$y,0,\bar{y}$ $y+\frac{1}{2},\frac{1}{4},y+\frac{3}{4}$ $\bar{y}+\frac{1}{4},\frac{3}{4},\bar{y}+\frac{1}{2}$ $\bar{y}+\frac{3}{4},\frac{1}{2},y+\frac{1}{4}$

$\bar{y},y,0$ $y+\frac{3}{4},y+\frac{1}{2},\frac{1}{4}$ $\bar{y}+\frac{1}{2},\bar{y}+\frac{1}{4},\frac{3}{4}$ $y+\frac{1}{4},\bar{y}+\frac{3}{4},\frac{1}{2}$

no extra conditions

96   g $..m$

x,x,z $\bar{x}+\frac{3}{4},\bar{x}+\frac{1}{4},z+\frac{1}{2}$ $\bar{x}+\frac{1}{4},x+\frac{1}{2},\bar{z}+\frac{3}{4}$ $x+\frac{1}{2},\bar{x}+\frac{3}{4},\bar{z}+\frac{1}{4}$

z,x,x $z+\frac{1}{2},\bar{x}+\frac{3}{4},\bar{x}+\frac{1}{4}$ $\bar{z}+\frac{3}{4},\bar{x}+\frac{1}{4},x+\frac{1}{2}$ $\bar{z}+\frac{1}{4},x+\frac{1}{2},\bar{x}+\frac{3}{4}$

x,z,x $\bar{x}+\frac{1}{4},z+\frac{1}{2},\bar{x}+\frac{3}{4}$ $x+\frac{1}{2},\bar{z}+\frac{3}{4},\bar{x}+\frac{1}{4}$ $\bar{x}+\frac{3}{4},\bar{z}+\frac{1}{4},x+\frac{1}{2}$

$x+\frac{3}{4},x+\frac{1}{4},\bar{z}+\frac{1}{2}$ $\bar{x},\bar{x},\bar{z}$ $x+\frac{1}{4},\bar{x}+\frac{1}{2},z+\frac{3}{4}$ $\bar{x}+\frac{1}{2},x+\frac{3}{4},z+\frac{1}{4}$

$x+\frac{3}{4},z+\frac{1}{4},\bar{x}+\frac{1}{2}$ $\bar{x}+\frac{1}{2},z+\frac{3}{4},x+\frac{1}{4}$ $\bar{x},\bar{z},\bar{x}$ $x+\frac{1}{4},\bar{z}+\frac{1}{2},x+\frac{3}{4}$

$z+\frac{3}{4},x+\frac{1}{4},\bar{x}+\frac{1}{2}$ $z+\frac{1}{4},\bar{x}+\frac{1}{2},x+\frac{3}{4}$ $\bar{z}+\frac{1}{2},x+\frac{3}{4},x+\frac{1}{4}$ $\bar{z},\bar{x},\bar{x}$

no extra conditions

48   f $2.mm$

$x,\frac{1}{8},\frac{1}{8}$ $\bar{x}+\frac{3}{4},\frac{1}{8},\frac{5}{8}$ $\frac{1}{8},x,\frac{1}{8}$ $\frac{5}{8},\bar{x}+\frac{3}{4},\frac{1}{8}$ $\frac{1}{8},\frac{1}{8},x$ $\frac{1}{8},\frac{5}{8},\bar{x}+\frac{3}{4}$

$\frac{7}{8},x+\frac{1}{4},\frac{3}{8}$ $\frac{7}{8},\bar{x},\frac{7}{8}$ $x+\frac{3}{4},\frac{3}{8},\frac{3}{8}$ $\bar{x}+\frac{1}{2},\frac{7}{8},\frac{3}{8}$ $\frac{7}{8},\frac{3}{8},\bar{x}+\frac{1}{2}$ $\frac{3}{8},\frac{3}{8},x+\frac{3}{4}$

hkl : $h=2n+1$
or $h+k+l=4n$

32   e $.3m$

x,x,x $\bar{x}+\frac{3}{4},\bar{x}+\frac{1}{4},x+\frac{1}{2}$

$\bar{x}+\frac{1}{4},x+\frac{1}{2},\bar{x}+\frac{3}{4}$ $x+\frac{1}{2},\bar{x}+\frac{3}{4},\bar{x}+\frac{1}{4}$

$x+\frac{3}{4},x+\frac{1}{4},\bar{x}+\frac{1}{2}$ $\bar{x},\bar{x},\bar{x}$

$x+\frac{1}{4},\bar{x}+\frac{1}{2},x+\frac{3}{4}$ $\bar{x}+\frac{1}{2},x+\frac{3}{4},x+\frac{1}{4}$

no extra conditions

16   d $.\bar{3}m$

$\frac{1}{2},\frac{1}{2},\frac{1}{2}$ $\frac{1}{4},\frac{3}{4},0$ $\frac{3}{4},0,\frac{1}{4}$ $0,\frac{1}{4},\frac{3}{4}$

16   c $.\bar{3}m$

$0,0,0$ $\frac{3}{4},\frac{1}{4},\frac{1}{2}$ $\frac{1}{4},\frac{1}{2},\frac{3}{4}$ $\frac{1}{2},\frac{3}{4},\frac{1}{4}$

hkl : $h=2n+1$
or $h,k,l=4n+2$
or $h,k,l=4n$

8   b $\bar{4}3m$

$\frac{3}{8},\frac{3}{8},\frac{3}{8}$ $\frac{1}{8},\frac{5}{8},\frac{1}{8}$

8   a $\bar{4}3m$

$\frac{1}{8},\frac{1}{8},\frac{1}{8}$ $\frac{7}{8},\frac{3}{8},\frac{3}{8}$

hkl : $h=2n+1$
or $h+k+l=4n$

Symmetry of special projections

Along [001] $p4mm$
$\mathbf{a}'=\frac{1}{4}(\mathbf{a}-\mathbf{b})$ $\mathbf{b}'=\frac{1}{4}(\mathbf{a}+\mathbf{b})$
Origin at $\frac{1}{8},\frac{3}{8},z$

Along [111] $p6mm$
$\mathbf{a}'=\frac{1}{6}(2\mathbf{a}-\mathbf{b}-\mathbf{c})$ $\mathbf{b}'=\frac{1}{6}(-\mathbf{a}+2\mathbf{b}-\mathbf{c})$
Origin at x,x,x

Along [110] $c2mm$
$\mathbf{a}'=\frac{1}{2}(-\mathbf{a}+\mathbf{b})$ $\mathbf{b}'=\mathbf{c}$
Origin at $x,x,0$

ORIGIN CHOICE 2

Maximal non-isomorphic subgroups

I   [2] $F \bar{4} 3 m$ (216)   (1; 2; 3; 4; 5; 6; 7; 8; 9; 10; 11; 12; 37; 38; 39; 40; 41; 42; 43; 44; 45; 46; 47; 48)+
       [2] $F 4_1 3 2$ (210)   (1; 2; 3; 4; 5; 6; 7; 8; 9; 10; 11; 12; 13; 14; 15; 16; 17; 18; 19; 20; 21; 22; 23; 24)+
       [2] $F d \bar{3} 1$ ($F d \bar{3}$, 203)   (1; 2; 3; 4; 5; 6; 7; 8; 9; 10; 11; 12; 25; 26; 27; 28; 29; 30; 31; 32; 33; 34; 35; 36)+
       ⎧ [3] $F 4_1/d 1 2/m$ ($I 4_1/a m d$, 141)   (1; 2; 3; 4; 13; 14; 15; 16; 25; 26; 27; 28; 37; 38; 39; 40)+
       ⎨ [3] $F 4_1/d 1 2/m$ ($I 4_1/a m d$, 141)   (1; 2; 3; 4; 17; 18; 19; 20; 25; 26; 27; 28; 41; 42; 43; 44)+
       ⎩ [3] $F 4_1/d 1 2/m$ ($I 4_1/a m d$, 141)   (1; 2; 3; 4; 21; 22; 23; 24; 25; 26; 27; 28; 45; 46; 47; 48)+
       ⎧ [4] $F 1 \bar{3} 2/m$ ($R \bar{3} m$, 166)   (1; 5; 9; 14; 19; 24; 25; 29; 33; 38; 43; 48)+
       ⎪ [4] $F 1 \bar{3} 2/m$ ($R \bar{3} m$, 166)   (1; 6; 12; 13; 18; 24; 25; 30; 36; 37; 42; 48)+
       ⎨ [4] $F 1 \bar{3} 2/m$ ($R \bar{3} m$, 166)   (1; 7; 10; 13; 19; 22; 25; 31; 34; 37; 43; 46)+
       ⎩ [4] $F 1 \bar{3} 2/m$ ($R \bar{3} m$, 166)   (1; 8; 11; 14; 18; 22; 25; 32; 35; 38; 42; 46)+

IIa   none
IIb   none

Maximal isomorphic subgroups of lowest index

IIc   [27] $F d \bar{3} m$ ($\mathbf{a}' = 3\mathbf{a}, \mathbf{b}' = 3\mathbf{b}, \mathbf{c}' = 3\mathbf{c}$) (227)

Minimal non-isomorphic supergroups

I   none
II   [2] $P n \bar{3} m$ ($\mathbf{a}' = \frac{1}{2}\mathbf{a}, \mathbf{b}' = \frac{1}{2}\mathbf{b}, \mathbf{c}' = \frac{1}{2}\mathbf{c}$) (224)

Symmetry operations

For $(0,0,0)+$ set

(1) 1	(2) $2(0,0,\frac{1}{2})\quad \frac{3}{8},\frac{1}{8},z$	(3) $2(0,\frac{1}{2},0)\quad \frac{1}{8},y,\frac{3}{8}$	(4) $2(\frac{1}{2},0,0)\quad x,\frac{3}{8},\frac{1}{8}$
(5) $3^{+}\ x,x,x$	(6) $3^{+}\ \bar{x}+\frac{1}{2},x+\frac{1}{4},\bar{x}$	(7) $3^{+}\ x+\frac{3}{4},\bar{x}-\frac{1}{2},\bar{x}$	(8) $3^{+}\ \bar{x}+\frac{1}{4},\bar{x}+\frac{3}{4},x$
(9) $3^{-}\ x,x,x$	(10) $3^{-}(-\frac{1}{3},\frac{1}{3},\frac{1}{3})\quad x+\frac{5}{12},\bar{x}+\frac{1}{6},\bar{x}$	(11) $3^{-}(\frac{1}{3},\frac{1}{3},-\frac{1}{3})\quad \bar{x}+\frac{7}{12},\bar{x}+\frac{5}{12},x$	(12) $3^{-}(\frac{1}{3},-\frac{1}{3},\frac{1}{3})\quad \bar{x}-\frac{1}{6},x+\frac{7}{12},\bar{x}$
(13) $2(\frac{1}{2},\frac{1}{2},0)\quad x,x-\frac{1}{4},\frac{1}{4}$	(14) $2\quad x,\bar{x},0$	(15) $4^{-}(0,0,\frac{3}{4})\quad \frac{3}{8},\frac{1}{8},z$	(16) $4^{+}(0,0,\frac{1}{4})\quad -\frac{1}{8},\frac{5}{8},z$
(17) $4^{-}(\frac{3}{4},0,0)\quad x,\frac{3}{8},\frac{1}{8}$	(18) $2(0,\frac{1}{2},\frac{1}{2})\quad \frac{1}{4},y+\frac{1}{4},y$	(19) $2\quad 0,y,\bar{y}$	(20) $4^{+}(\frac{1}{4},0,0)\quad x,-\frac{1}{8},\frac{5}{8}$
(21) $4^{+}(0,\frac{1}{4},0)\quad \frac{5}{8},y,-\frac{1}{8}$	(22) $2(\frac{1}{2},0,\frac{1}{2})\quad x-\frac{1}{4},\frac{1}{4},x$	(23) $4^{-}(0,\frac{3}{4},0)\quad \frac{1}{8},y,\frac{3}{8}$	(24) $2\quad \bar{x},0,x$
(25) $\bar{1}\quad 0,0,0$	(26) $d(\frac{1}{4},\frac{3}{4},0)\quad x,y,\frac{1}{4}$	(27) $d(\frac{3}{4},0,\frac{1}{4})\quad x,\frac{1}{4},z$	(28) $d(0,\frac{1}{4},\frac{3}{4})\quad \frac{1}{4},y,z$
(29) $\bar{3}^{+}\ x,x,x;\ 0,0,0$	(30) $\bar{3}^{+}\ \bar{x}-1,x+\frac{3}{4},\bar{x};\ -\frac{1}{4},0,\frac{3}{4}$	(31) $\bar{3}^{+}\ x-\frac{1}{4},\bar{x}+1,\bar{x};\ 0,\frac{3}{4},-\frac{1}{4}$	(32) $\bar{3}^{+}\ \bar{x}+\frac{3}{4},\bar{x}-\frac{1}{4},x;\ \frac{3}{4},-\frac{1}{4},0$
(33) $\bar{3}^{-}\ x,x,x;\ 0,0,0$	(34) $\bar{3}^{-}\ x+\frac{5}{4},\bar{x}-1,\bar{x};\ \frac{1}{2},-\frac{1}{4},\frac{3}{4}$	(35) $\bar{3}^{-}\ \bar{x}+\frac{1}{4},\bar{x}+\frac{5}{4},x;\ -\frac{1}{4},\frac{3}{4},\frac{1}{2}$	(36) $\bar{3}^{-}\ \bar{x}+1,x+\frac{1}{4},\bar{x};\ \frac{3}{4},\frac{1}{2},-\frac{1}{4}$
(37) $g(-\frac{1}{4},\frac{1}{4},\frac{1}{2})\quad x+\frac{1}{2},\bar{x},z$	(38) $m\quad x,x,z$	(39) $\bar{4}^{-}\ \frac{1}{8},\frac{5}{8},z;\ \frac{1}{8},\frac{5}{8},\frac{1}{8}$	(40) $\bar{4}^{+}\ \frac{3}{8},-\frac{1}{8},z;\ \frac{3}{8},-\frac{1}{8},\frac{3}{8}$
(41) $\bar{4}^{-}\ x,\frac{1}{8},\frac{5}{8};\ \frac{1}{8},\frac{1}{8},\frac{5}{8}$	(42) $g(\frac{1}{2},-\frac{1}{4},\frac{1}{4})\quad x,y+\frac{1}{2},\bar{y}$	(43) $m\quad x,y,y$	(44) $\bar{4}^{+}\ x,\frac{3}{8},-\frac{1}{8};\ \frac{3}{8},\frac{3}{8},-\frac{1}{8}$
(45) $\bar{4}^{+}\ -\frac{1}{8},y,\frac{3}{8};\ -\frac{1}{8},\frac{3}{8},\frac{3}{8}$	(46) $g(\frac{1}{4},\frac{1}{2},-\frac{1}{4})\quad \bar{x}+\frac{1}{2},y,x$	(47) $\bar{4}^{-}\ \frac{5}{8},y,\frac{1}{8};\ \frac{5}{8},\frac{1}{8},\frac{1}{8}$	(48) $m\quad x,y,x$

For $(0,\frac{1}{2},\frac{1}{2})+$ set

(1) $t(0,\frac{1}{2},\frac{1}{2})$	(2) $2\quad \frac{3}{8},\frac{3}{8},z$	(3) $2\quad \frac{1}{8},y,\frac{1}{8}$	(4) $2(\frac{1}{2},0,0)\quad x,\frac{1}{8},\frac{3}{8}$
(5) $3^{+}(\frac{1}{3},\frac{1}{3},\frac{1}{3})\quad x-\frac{1}{3},x-\frac{1}{6},x$	(6) $3^{+}(\frac{1}{3},-\frac{1}{3},\frac{1}{3})\quad \bar{x}+\frac{1}{6},x+\frac{5}{12},\bar{x}$	(7) $3^{+}\ x+\frac{3}{4},\bar{x},\bar{x}$	(8) $3^{+}\ \bar{x}+\frac{1}{4},\bar{x}+\frac{1}{4},x$
(9) $3^{-}(\frac{1}{3},\frac{1}{3},\frac{1}{3})\quad x-\frac{1}{6},x+\frac{1}{6},x$	(10) $3^{-}\ x+\frac{1}{4},\bar{x},\bar{x}$	(11) $3^{-}\ \bar{x}+\frac{3}{4},\bar{x}+\frac{1}{4},x$	(12) $3^{-}\ \bar{x},x+\frac{3}{4},\bar{x}$
(13) $2(\frac{3}{4},\frac{3}{4},0)\quad x,x,0$	(14) $2(-\frac{1}{4},\frac{1}{4},0)\quad x,\bar{x}+\frac{1}{4},\frac{1}{4}$	(15) $4^{-}(0,0,\frac{1}{4})\quad \frac{1}{8},-\frac{1}{8},z$	(16) $4^{+}(0,0,\frac{3}{4})\quad \frac{1}{8},\frac{3}{8},z$
(17) $4^{-}(\frac{3}{4},0,0)\quad x,\frac{3}{8},-\frac{3}{8}$	(18) $2(0,\frac{1}{2},\frac{1}{2})\quad \frac{1}{4},y-\frac{1}{4},y$	(19) $2\quad 0,y+\frac{1}{2},\bar{y}$	(20) $4^{+}(\frac{1}{4},0,0)\quad x,-\frac{1}{8},\frac{1}{8}$
(21) $4^{+}(0,\frac{3}{4},0)\quad \frac{3}{8},y,-\frac{3}{8}$	(22) $2(\frac{1}{4},0,\frac{1}{4})\quad x,0,x$	(23) $4^{-}(0,\frac{1}{4},0)\quad -\frac{1}{8},y,\frac{5}{8}$	(24) $2(-\frac{1}{4},0,\frac{1}{4})\quad \bar{x}+\frac{1}{4},\frac{1}{4},x$
(25) $\bar{1}\quad 0,\frac{1}{4},\frac{1}{4}$	(26) $d(\frac{1}{4},\frac{1}{4},0)\quad x,y,0$	(27) $d(\frac{3}{4},0,\frac{3}{4})\quad x,0,z$	(28) $d(0,\frac{3}{4},\frac{1}{4})\quad \frac{1}{4},y,z$
(29) $\bar{3}^{+}\ x,x+\frac{1}{2},x;\ 0,\frac{1}{2},0$	(30) $\bar{3}^{+}\ \bar{x}-1,x+\frac{5}{4},\bar{x};\ -\frac{1}{4},\frac{1}{2},\frac{3}{4}$	(31) $\bar{3}^{+}\ x-\frac{1}{4},\bar{x}+\frac{1}{2},\bar{x};\ 0,\frac{1}{4},-\frac{1}{4}$	(32) $\bar{3}^{+}\ \bar{x}+\frac{3}{4},\bar{x}-\frac{3}{4},x;\ \frac{3}{4},-\frac{3}{4},0$
(33) $\bar{3}^{-}\ x-\frac{1}{2},x-\frac{1}{2},x;\ 0,0,\frac{1}{2}$	(34) $\bar{3}^{-}\ x+\frac{3}{4},\bar{x}-\frac{3}{2},\bar{x};\ 0,-\frac{3}{4},\frac{3}{4}$	(35) $\bar{3}^{-}\ \bar{x}-\frac{1}{4},\bar{x}+\frac{3}{4},x;\ -\frac{1}{4},\frac{3}{4},0$	(36) $\bar{3}^{-}\ \bar{x}+\frac{1}{2},x-\frac{1}{4},\bar{x};\ \frac{1}{4},0,-\frac{1}{4}$
(37) $m\quad x+\frac{1}{4},\bar{x},z$	(38) $g(\frac{1}{4},\frac{1}{4},\frac{1}{2})\quad x-\frac{1}{4},x,z$	(39) $\bar{4}^{-}\ \frac{3}{8},\frac{3}{8},z;\ \frac{3}{8},\frac{3}{8},\frac{3}{8}$	(40) $\bar{4}^{+}\ \frac{5}{8},\frac{1}{8},z;\ \frac{5}{8},\frac{1}{8},\frac{1}{8}$
(41) $\bar{4}^{-}\ x,\frac{1}{8},\frac{1}{8};\ \frac{1}{8},\frac{1}{8},\frac{1}{8}$	(42) $g(\frac{1}{2},\frac{1}{4},-\frac{1}{4})\quad x,y+\frac{1}{2},\bar{y}$	(43) $g(0,\frac{1}{2},\frac{1}{2})\quad x,y,y$	(44) $\bar{4}^{+}\ x,\frac{3}{8},\frac{3}{8};\ \frac{3}{8},\frac{3}{8},\frac{3}{8}$
(45) $\bar{4}^{+}\ \frac{1}{8},y,\frac{1}{8};\ \frac{1}{8},\frac{1}{8},\frac{1}{8}$	(46) $m\quad \bar{x}+\frac{3}{4},y,x$	(47) $\bar{4}^{-}\ \frac{3}{8},y,-\frac{1}{8};\ \frac{3}{8},\frac{3}{8},-\frac{1}{8}$	(48) $g(\frac{1}{4},\frac{1}{2},\frac{1}{4})\quad x-\frac{1}{4},y,x$

For $(\frac{1}{2},0,\frac{1}{2})+$ set

(1) $t(\frac{1}{2},0,\frac{1}{2})$	(2) $2\quad \frac{1}{8},\frac{1}{8},z$	(3) $2(0,\frac{1}{2},0)\quad \frac{3}{8},y,\frac{1}{8}$	(4) $2\quad x,\frac{3}{8},\frac{3}{8}$
(5) $3^{+}(\frac{1}{3},\frac{1}{3},\frac{1}{3})\quad x+\frac{1}{6},x-\frac{1}{6},x$	(6) $3^{+}\ \bar{x},x+\frac{3}{4},\bar{x}$	(7) $3^{+}\ x+\frac{1}{4},\bar{x},\bar{x}$	(8) $3^{+}(\frac{1}{3},\frac{1}{3},-\frac{1}{3})\quad \bar{x}+\frac{5}{12},\bar{x}+\frac{7}{12},x$
(9) $3^{-}(\frac{1}{3},\frac{1}{3},\frac{1}{3})\quad x-\frac{1}{6},x-\frac{1}{3},x$	(10) $3^{-}\ x+\frac{1}{4},\bar{x}+\frac{1}{2},\bar{x}$	(11) $3^{-}\ \bar{x}+\frac{3}{4},\bar{x}+\frac{3}{4},x$	(12) $3^{-}\ \bar{x},x+\frac{1}{4},\bar{x}$
(13) $2(\frac{1}{4},\frac{1}{4},0)\quad x,x,0$	(14) $2(\frac{1}{4},-\frac{1}{4},0)\quad x,\bar{x}+\frac{1}{4},\frac{1}{4}$	(15) $4^{-}(0,0,\frac{1}{4})\quad \frac{5}{8},-\frac{1}{8},z$	(16) $4^{+}(0,0,\frac{3}{4})\quad -\frac{3}{8},\frac{3}{8},z$
(17) $4^{-}(\frac{1}{4},0,0)\quad x,\frac{1}{8},-\frac{1}{8}$	(18) $2(0,\frac{3}{4},\frac{3}{4})\quad 0,y,y$	(19) $2(0,-\frac{1}{4},\frac{1}{4})\quad \frac{1}{4},y+\frac{1}{4},\bar{y}$	(20) $4^{+}(\frac{3}{4},0,0)\quad x,\frac{1}{8},\frac{3}{8}$
(21) $4^{+}(0,\frac{1}{4},0)\quad \frac{1}{8},y,-\frac{1}{8}$	(22) $2(\frac{1}{2},0,\frac{1}{2})\quad x+\frac{1}{4},\frac{1}{4},x$	(23) $4^{-}(0,\frac{3}{4},0)\quad -\frac{3}{8},y,\frac{3}{8}$	(24) $2\quad \bar{x}+\frac{1}{2},0,x$
(25) $\bar{1}\quad \frac{1}{4},0,\frac{1}{4}$	(26) $d(\frac{3}{4},\frac{3}{4},0)\quad x,y,0$	(27) $d(\frac{1}{4},0,\frac{3}{4})\quad x,\frac{1}{4},z$	(28) $d(0,\frac{1}{4},\frac{1}{4})\quad 0,y,z$
(29) $\bar{3}^{+}\ x-\frac{1}{2},x-\frac{1}{2},x;\ 0,0,\frac{1}{2}$	(30) $\bar{3}^{+}\ \bar{x}-\frac{1}{2},x+\frac{1}{4},\bar{x};\ -\frac{1}{4},0,\frac{1}{4}$	(31) $\bar{3}^{+}\ x-\frac{3}{4},\bar{x}+\frac{3}{2},\bar{x};\ 0,\frac{3}{4},-\frac{3}{4}$	(32) $\bar{3}^{+}\ \bar{x}+\frac{5}{4},\bar{x}+\frac{1}{4},x;\ \frac{3}{4},-\frac{1}{4},\frac{1}{2}$
(33) $\bar{3}^{-}\ x+\frac{1}{2},x,x;\ \frac{1}{2},0,0$	(34) $\bar{3}^{-}\ x+\frac{3}{4},\bar{x}-1,\bar{x};\ 0,-\frac{1}{4},\frac{3}{4}$	(35) $\bar{3}^{-}\ \bar{x}-\frac{1}{4},\bar{x}+\frac{1}{4},x;\ -\frac{1}{4},\frac{1}{4},0$	(36) $\bar{3}^{-}\ \bar{x}+\frac{3}{2},x-\frac{3}{4},\bar{x};\ \frac{3}{4},0,-\frac{3}{4}$
(37) $m\quad x+\frac{3}{4},\bar{x},z$	(38) $g(\frac{1}{4},\frac{1}{4},\frac{1}{2})\quad x+\frac{1}{4},x,z$	(39) $\bar{4}^{-}\ -\frac{1}{8},\frac{3}{8},z;\ -\frac{1}{8},\frac{3}{8},\frac{3}{8}$	(40) $\bar{4}^{+}\ \frac{1}{8},\frac{1}{8},z;\ \frac{1}{8},\frac{1}{8},\frac{1}{8}$
(41) $\bar{4}^{-}\ x,\frac{3}{8},\frac{3}{8};\ \frac{3}{8},\frac{3}{8},\frac{3}{8}$	(42) $m\quad x,y+\frac{1}{4},\bar{y}$	(43) $g(\frac{1}{2},\frac{1}{4},\frac{1}{4})\quad x,y-\frac{1}{4},y$	(44) $\bar{4}^{+}\ x,\frac{5}{8},\frac{1}{8};\ \frac{1}{8},\frac{5}{8},\frac{1}{8}$
(45) $\bar{4}^{+}\ \frac{3}{8},y,\frac{3}{8};\ \frac{3}{8},\frac{3}{8},\frac{3}{8}$	(46) $g(-\frac{1}{4},\frac{1}{2},\frac{1}{4})\quad \bar{x}+\frac{1}{2},y,x$	(47) $\bar{4}^{-}\ \frac{1}{8},y,\frac{1}{8};\ \frac{1}{8},\frac{1}{8},\frac{1}{8}$	(48) $g(\frac{1}{2},0,\frac{1}{2})\quad x,y,x$

For $(\frac{1}{2},\frac{1}{2},0)+$ set

(1) $t(\frac{1}{2},\frac{1}{2},0)$	(2) $2(0,0,\frac{1}{2})\quad \frac{1}{8},\frac{3}{8},z$	(3) $2\quad \frac{3}{8},y,\frac{3}{8}$	(4) $2\quad x,\frac{1}{8},\frac{1}{8}$
(5) $3^{+}(\frac{1}{3},\frac{1}{3},\frac{1}{3})\quad x+\frac{1}{6},x+\frac{1}{3},x$	(6) $3^{+}\ \bar{x},x+\frac{1}{4},\bar{x}$	(7) $3^{+}(-\frac{1}{3},\frac{1}{3},\frac{1}{3})\quad x+\frac{7}{12},\bar{x}-\frac{1}{6},\bar{x}$	(8) $3^{+}\ \bar{x}+\frac{3}{4},\bar{x}+\frac{3}{4},x$
(9) $3^{-}(\frac{1}{3},\frac{1}{3},\frac{1}{3})\quad x+\frac{1}{3},x+\frac{1}{6},x$	(10) $3^{-}\ x+\frac{3}{4},\bar{x},\bar{x}$	(11) $3^{-}\ \bar{x}+\frac{1}{4},\bar{x}+\frac{1}{4},x$	(12) $3^{-}\ \bar{x}-\frac{1}{2},x+\frac{3}{4},\bar{x}$
(13) $2(\frac{1}{2},\frac{1}{2},0)\quad x,x+\frac{1}{4},\frac{1}{4}$	(14) $2\quad x,\bar{x}+\frac{1}{2},0$	(15) $4^{-}(0,0,\frac{3}{4})\quad \frac{3}{8},-\frac{3}{8},z$	(16) $4^{+}(0,0,\frac{1}{4})\quad -\frac{1}{8},\frac{1}{8},z$
(17) $4^{-}(\frac{1}{4},0,0)\quad x,\frac{5}{8},-\frac{1}{8}$	(18) $2(0,\frac{1}{4},\frac{1}{4})\quad 0,y,y$	(19) $2(0,\frac{1}{4},-\frac{1}{4})\quad \frac{1}{4},y+\frac{1}{4},\bar{y}$	(20) $4^{+}(\frac{3}{4},0,0)\quad x,-\frac{3}{8},\frac{3}{8}$
(21) $4^{+}(0,\frac{3}{4},0)\quad \frac{3}{8},y,\frac{1}{8}$	(22) $2(\frac{3}{4},0,\frac{3}{4})\quad x,0,x$	(23) $4^{-}(0,\frac{1}{4},0)\quad -\frac{1}{8},y,\frac{1}{8}$	(24) $2(\frac{1}{4},0,-\frac{1}{4})\quad \bar{x}+\frac{1}{4},\frac{1}{4},x$
(25) $\bar{1}\quad \frac{1}{4},\frac{1}{4},0$	(26) $d(\frac{3}{4},\frac{1}{4},0)\quad x,y,\frac{1}{4}$	(27) $d(\frac{1}{4},0,\frac{1}{4})\quad x,0,z$	(28) $d(0,\frac{3}{4},\frac{3}{4})\quad 0,y,z$
(29) $\bar{3}^{+}\ x+\frac{1}{2},x,x;\ \frac{1}{2},0,0$	(30) $\bar{3}^{+}\ \bar{x}-\frac{3}{2},x+\frac{3}{4},\bar{x};\ -\frac{3}{4},0,\frac{3}{4}$	(31) $\bar{3}^{+}\ x+\frac{1}{4},\bar{x}+1,\bar{x};\ \frac{1}{2},\frac{3}{4},-\frac{1}{4}$	(32) $\bar{3}^{+}\ \bar{x}+\frac{1}{4},\bar{x}-\frac{1}{4},x;\ \frac{1}{4},-\frac{1}{4},0$
(33) $\bar{3}^{-}\ x,x+\frac{1}{2},x;\ 0,\frac{1}{2},0$	(34) $\bar{3}^{-}\ x+\frac{1}{4},\bar{x}-\frac{1}{2},\bar{x};\ 0,-\frac{1}{4},\frac{1}{4}$	(35) $\bar{3}^{-}\ \bar{x}-\frac{3}{4},\bar{x}+\frac{3}{4},x;\ -\frac{3}{4},\frac{3}{4},0$	(36) $\bar{3}^{-}\ \bar{x}+1,x-\frac{1}{4},\bar{x};\ \frac{3}{4},0,-\frac{1}{4}$
(37) $g(\frac{1}{4},-\frac{1}{4},\frac{1}{2})\quad x+\frac{1}{2},\bar{x},z$	(38) $g(\frac{1}{2},\frac{1}{2},0)\quad x,x,z$	(39) $\bar{4}^{-}\ \frac{1}{8},\frac{1}{8},z;\ \frac{1}{8},\frac{1}{8},\frac{1}{8}$	(40) $\bar{4}^{+}\ \frac{3}{8},\frac{3}{8},z;\ \frac{3}{8},\frac{3}{8},\frac{3}{8}$
(41) $\bar{4}^{-}\ x,-\frac{1}{8},\frac{3}{8};\ \frac{3}{8},-\frac{1}{8},\frac{3}{8}$	(42) $m\quad x,y+\frac{3}{4},\bar{y}$	(43) $g(\frac{1}{2},\frac{1}{4},\frac{1}{4})\quad x,y+\frac{1}{4},y$	(44) $\bar{4}^{+}\ x,\frac{1}{8},\frac{1}{8};\ \frac{1}{8},\frac{1}{8},\frac{1}{8}$
(45) $\bar{4}^{+}\ \frac{1}{8},y,\frac{5}{8};\ \frac{1}{8},\frac{1}{8},\frac{5}{8}$	(46) $m\quad \bar{x}+\frac{1}{4},y,x$	(47) $\bar{4}^{-}\ \frac{3}{8},y,\frac{3}{8};\ \frac{3}{8},\frac{3}{8},\frac{3}{8}$	(48) $g(\frac{1}{4},\frac{1}{2},\frac{1}{4})\quad x+\frac{1}{4},y,x$

Author index

Entries refer to chapter number.

162

Subject index

① *Headline* in abbreviated form.

② *Generators selected*: Sections 2.2.10 and 8.3.5. A set of generators, as selected for these *Tables*, is listed in the form of translations and numbers of general-position coordinates. The generators determine the sequence of the coordinate triplets in the general position and of the corresponding symmetry operations.

③ *Positions*: Sections 2.2.11 and 8.3.2. The general Wyckoff position is given at the top, followed downwards by the various special Wyckoff positions with decreasing multiplicity and increasing site symmetry. For each general and special position its multiplicity, Wyckoff letter, oriented site-symmetry symbol, as well as the appropriate coordinate triplets and the reflection conditions, are listed. The coordinate triplets of the general position are numbered sequentially; *cf. Symmetry operations*.

Oriented site-symmetry symbol (third column): Section 2.2.12. The site symmetry at the points of a special position is given in oriented form.

Reflection conditions (right-most column): Section 2.2.13.

[*Lattice complexes* are described in Part 14; Tables 14.2.3.1 and 14.2.3.2 show the assignment of Wyckoff positions to Wyckoff sets and to lattice complexes.]

④ *Symmetry of special projections*: Section 2.2.14. For each space group, orthographic projections along three (symmetry) directions are listed. Given are the projection direction, the plane group of the projection, as well as the axes and the origin of the projected cell.

⑤ *Maximal non-isomorphic subgroups*: Sections 2.2.15 and 8.3.3.

Type **I**: *translationengleiche* or *t* subgroups;
Type **IIa**: *klassengleiche* or *k* subgroups, obtained by 'decentring' the conventional cell; applies only to space groups with centred cells;
Type **IIb**: *klassengleiche* or *k* subgroups, obtained by enlarging the conventional cell.

Given are:
   For types **I** and **IIa**: Index [between brackets]; 'unconventional' Hermann–Mauguin symbol of the subgroup; 'conventional' Hermann–Mauguin symbol of the subgroup, if different (between parentheses); coordinate triplets retained in subgroup.
   For type **IIb**: Index [between brackets]; 'unconventional' Hermann–Mauguin symbol of the subgroup; basis-vector relations between group and subgroup (between parentheses); 'conventional' Hermann–Mauguin symbol of the subgroup, if different (between parentheses).

⑥ *Maximal isomorphic subgroups of lowest index*: Sections 2.2.15, 8.3.3 and 13.1.2.

Type **IIc**: *klassengleiche* or *k* subgroups of lowest index which are of the same type as the group, *i.e.* have the same standard Hermann–Mauguin symbol. Data as for subgroups of type **IIb**.

⑦ *Minimal non-isomorphic supergroups*: Sections 2.2.15 and 8.3.3.
   The list contains the reverse relations of the subgroup tables; only types **I** (*t* supergroups) and **II** (*k* supergroups) are distinguished. Data as for subgroups of type **IIb**.

EXPLANATION OF THE SPACE-GROUP DATA